ENVIRONMENTAL HAZARDS

AIR POLLUTION

ENVIRONMENTAL HAZARDS

AIR POLLUTION

A Reference Handbook

E. Willard Miller
Department of Geography

Ruby M. Miller
Pattee Library

The Pennsylvania State University

CONTEMPORARY WORLD ISSUES

ABC-CLIO

Santa Barbara, California
Oxford, England

© 1989 by ABC-CLIO, Inc.

Library of Congress Cataloging-in-Publication Data

Miller, E. Willard (Eugene Willard), 1915–
    Environmental hazards : air pollution : a reference handbook
/ E. Willard Miller, Ruby M. Miller.
(Contemporary world issues)
    Bibliography: p.
    Includes index.
    1. Air—Pollution—United States. 2. Air—Pollution—Law and legislation—United States. 3. Environmental policy—United States. I. Miller, Ruby M. II. Title. III. Series: Contemporary world issues series.
HC110.A4M55 1989   363.7′392′0973—dc19   88-39233

ISBN 0-87436-528-7 (alk. paper)

96 95 94 93 92 91   10 9 8 7 6 5 4 3

ABC-CLIO, Inc.
130 Cremona, P.O. Box 1911
Santa Barbara, California 93116-1911

Clio Press Ltd.
55 St. Thomas' Street
Oxford, OX1 1JG, England

This book is Smyth-sewn and printed on acid-free paper ∞.
Manufactured in the United States of America.

Contents

Preface

THE ATMOSPHERE IS one of the earth's most precious resources, and until recently it was thought that the atmosphere was so vast that it could not be affected by human activity. With the coming of the age of environmental awareness and concern, issues of air quality are looming as major public policy frontiers in the 1990s, and we are now aware that human activities can alter the world's atmosphere.

The problem of pollution of the atmosphere is not a new one, but it has reached a critical dimension in our time. Modern wastes spewed into the atmosphere have the potential of not only changing the natural environment but affecting the health of millions of human beings.

Because of the vastness of the atmosphere, the problems of air pollution are created over decades, and it is difficult, if not impossible, to detect short-range alterations. As a result, policy decisions must be made on a long-range basis. For example, it is now known that the carbon dioxide content of the atmosphere began to increase after the Industrial Revolution began a century or so ago, and questions are now being raised as to the consequences of this worldwide change in the content of the atmosphere. A fundamental question is, Has the increase in the carbon dioxide content of the atmosphere caused climatic changes or will there be changes in the future? Although such questions cannot be answered definitely at present, they should not be ignored, for models indicate that great potential changes could occur when a critical threshold has been reached. To alter the world's energy system will create vast economic changes, but there is a need to develop energy systems that will not pollute the atmosphere, including increased use of solar energy, wind power, and energy from the oceans and the development of safe nuclear power.

This volume begins with a description and analysis of the broad aspects of air pollution. A discussion of natural air pollution is followed by sections on environmental hazards that are caused by atmospheric pollution. The most important of the pollutants are sulfur dioxide, carbon dioxide, nitric oxides, chloro-fluorocarbons, radioactive emissions, and radon. These pollutants provide the basis for the creation of acid precipitation, the greenhouse effect, the depletion of ozone in the stratosphere, and potential deleterious effects on human health.

As the problems of air pollution have become evident, legislation has evolved in order to protect the general public, and Chapter 3 discusses the development of legislation in the United States from the 1940s to the present. The ineffectiveness of voluntary local and state efforts made the control of air pollutants a prominent national concern, and the Clean Air Act Amendments of 1970 marked the first comprehensive attempt to provide air standards for the nation. Although legislation has become increasingly comprehensive, it is now evident that attaining clean air is a much more difficult task than originally visualized.

Other chapters provide information on a variety of topics. A chronology lists some of the critical dates and shows that atmospheric hazards have plagued humankind for centuries, and short biographies are presented for a number of individuals who have made significant contributions to solving the problem of air pollution. Many organizations have been established to consider the problems of air pollution, and an organizational directory is divided into three major parts. The first considers the private organizations that represent a discipline, such as the American Meteorological Society; the second lists U.S. government agencies, of which the Environmental Protection Agency is the major one; and the third lists international organizations.

In recent years there has been a massive increase in the amount of literature on atmospheric pollution as a response not only to the potential danger to the physical and biological environment but also to its effects on human health. In Chapter 5, over 100˙ books are briefly annotated, and approximately 390 journal articles and government documents are listed. The references are topically arranged, and at the end of the bibliography there is a list of selected journals that publish articles on the atmospheric environment. The volume concludes with an annotated list of films and film strips dealing with environmental problems and a glossary.

E. Willard Miller
Ruby M. Miller
The Pennsylvania State University

Air Pollution: A Perspective

ALTHOUGH AIR POLLUTION IS RECOGNIZED as one of the most important, controversial environmental dilemmas of the modern day, it is also one of the world's oldest problems. There is evidence that when humans first built fires in caves, smoke filled the living quarters, and even when crude houses were constructed, there were no chimneys, and smoke filled the rooms; some primitive cultures still suffer from this problem today. After its invention, the chimney removed the unburned particles of combustion from the room, and the quality of indoor air improved greatly.

In the Bronze and Iron ages, industrial villages thrived and were sources of smoke and particulates in the atmosphere. Clay was baked and glazed to form pottery as early as 4000 B.C., and native copper and gold were also forged at a very early date. Iron was in common use before 1000 B.C., and most of the methods of modern metallurgy were known before A.D. 1. The early industries relied on charcoal, and the conversion of wood to charcoal created a smoky environment. Coal was used before A.D. 1000 and began to be converted into coke about 1600.

Over the centuries there have also been references in literature to the problems of smoke in the cities. As early as A.D. 61 the Roman philosopher Seneca wrote about pollution in Rome, "As soon as I had gotten out of the heavy air of Rome and from the stink of the smoky chimneys thereof, which, being stirred, poured forth whatever pestilential vapors and soot they had enclosed in them, I felt an alteration of my disposition." The problem of pollution in London has been reported for centuries. Conditions became so bad that in 1306 Edward I issued a royal proclamation forbidding the use of "sea coal" in the city, and Queen Elizabeth I banned the burning of coal in London when Parliament was in session. In 1661 John Evelyn issued the brochure *Fumifugium, or the Inconvenience of Aer and Smoake of London Dissipated* for the edification of King Charles II and Parliament. The brochure outlined the problem and offered suggestions as to how to remedy the situation. Later parliamentary actions (see Chapter 1) are proof that the problem in London did not abate for some time.

With the coming of the Industrial Revolution, which began in England in the middle of the eighteenth century, air pollution increased tremendously. The most important fuel of this revolution was coal, and in the nineteenth century petroleum and natural gas gradually became other major sources of energy. The predominant air pollution problem of this era was caused by smoke and particulates from the burning of coal or oil in the furnaces of the power plants, in locomotives, and in home fireplaces and furnaces. Because the situation initially was worse in Britain, the British were leaders in investigating the problem and in having some limited success in controlling air pollution.

In the volume *Problems and Control of Air Pollution,* Sir Hugh E.C. Beaver wrote:

By 1819, there was sufficient pressure for Parliament to appoint the first of a whole dynasty of committees "to consider how far persons using steam engines and furnaces could work them in a manner less prejudicial to public health and comfort." This committee confirmed the practicability of smoke prevention, as so many succeeding committees were to do, but as was often again to be experienced, nothing was done.

In 1843, there was another Parliamentary Select Committee, and in 1845, a third. In the same year, during the height of the great railway boom, an act of Parliament disposed of trouble from locomotives once and for all [!] by laying down the dictum that they must consume their own smoke. The Town Improvement Clauses Act two years later applied the same panacea to factory furnaces. Then 1853 and 1856 witnessed two acts of Parliament dealing specifically with London and empowering the police to enforce provisions against smoke from furnaces, public baths and washhouses and furnaces used in the working of steam vessels on the Thames. [New York: Reinhold Publishing Corporation, 1955, p. 3]

In the United Kingdom smoke and ash pollution was considered a health problem, and public health acts were passed in 1848, 1866, and 1875 to establish controls. Air pollution from the chemical industries was considered separately, and controls were established in the Alkali Act of 1863.

In the United States, control of air pollution was long considered to be, not a state or federal function, but the responsibility of municipal governments. The first municipal regulations were passed in the 1880s, but these regulations were ineffective, and by 1925 air pollution in urban and industrial centers had risen dramatically. Economic welfare was considered more important than smoke and ash abatement, and a polluted atmosphere indicated that employment was high and prosperity prevailed.

By 1925 air pollution was universal in all industrial nations, and there was a growing recognition that this situation was intolerable. As a first step toward finding a solution, large-scale surveys were made of pollution in cities—Salt Lake City in 1926, New York City in 1937, and Leicester, England, in 1939. Because of the widespread pollution by motor vehicles, air pollution research centered in California. The first National Air Pollution Symposium in the United States was held in Pasadena, California, in 1949, and the first U.S. Technical Conference on Air Pollution was held in Washington, D.C., in 1950. The technical foundation was being laid for the study of the dissemination of atmospheric pollutants, but no significant natural air pollution control legislation or regulation was immediately passed.

After 1945, some major air pollution incidents—such as those in Donora, Pennsylvania, and London—were catalysts for the development of legislation. One of the earliest acts was the Clean Air Act of 1956 in Britain, and by 1970, not only Britain but also the United States, most European countries, Japan, Australia, and New Zealand had enacted national air pollution control legislation. Air pollution research also expanded tremendously. In 1959 the first International Air Pollution Conference was held in London, and three national air pollution conferences were held in Washington, D.C., in 1958, 1962, and 1966. By 1970 major national air pollution centers had been established in the United States, Britain, France, East Germany, West Germany, the Netherlands, Sweden, and Japan.

Also by the 1970s mathematical models of air pollution were being developed. Air quality monitoring systems were being established all over the world, and measuring instruments had become sophisticated. Most important was the emergence of the ecological, or total environmental, approach, which involved not only scientific advancements but also political organization. National departments were established, and in the United States, the Environmental Protection Agency (EPA) has the responsibility of administering the protective legislation. In addition, most states have established counterpart organizations charged with the responsibility of maintaining a clean environment.

What Is Clean Air?

In attempting to define air pollution, the major question to be asked is, What constitutes clean air? If clean air is defined as that found far removed from human habitation and not affected by catastrophic and violent natural phenomena, a sample of such air would contain many substances other than oxygen and nitrogen, its major constituents. Rare gases such as argon, neon, and helium would be present, as well as ozone, carbon dioxide, radioactive materials from the earth, and various nitrogen and sulfur compounds. This air sample would also contain variable amounts of water vapor and a great many suspended solid particles and liquid substances.

The suspended material known collectively as aerosols may be defined as dust particles and condensation nuclei. The latter consist of chloride salts, sulfuric and nitrous acids, phosphorus compounds, and many other chemical substances. These nuclei have an affinity for water and thus play an important role in the transformation of water vapor into fog, clouds, and precipitation. In contrast, the dust particles are ordinarily inactive with respect to the condensation of water vapor. Another distinctive feature of dust particles is that the average particle is about ten times larger than the average nucleus. The number of nuclei in clean air outnumbers the number of dust particles many times over, and because of their small size, the nuclei are easily transported by air currents to higher altitudes and far greater distances than are dust particles.

Geographic Sources of Air Pollution

three geographic sources of air pollution: point source,
nd regional source. The point source of pollution is a
emissions into the atmosphere. The pollution may
emission, as in a manufacturing process, or as a
short period of time, as in a nuclear or volcanic
tter case, the distribution of the pollutants will
e of the explosion and the wind speed and
of emission. If the explosion penetrates the
lution can be worldwide and extend over
explosion is limited to the troposphere, the
more limited. If the air is calm, the pollutant
ed in a small area, but wind movement will
r a wider area. The steady point sources of
the lower layers of the atmosphere are the most
most evident to the general public, and the most

A line source of pollutants occurs when a number of point
sources are connected. A heavily trafficked highway is a line
source of pollutants, as are a number of chemical or heavy
industrial plants strung out along a valley route. If the line source
is long enough, the dispersion of pollution will occur both vertically and horizontally. Dispersal along a highway can become a
major problem if the wind flows parallel to the highway because in
this situation, the pollutants may be concentrated at a particular
place downwind.

Regional sources may vary in size. Atmospheric pollution
may be limited to a few square miles over an industrial park, or it
may cover a million square miles. Regional sources are made up of
many point and line sources that combine to pollute a large area,
and these sources may include all types of atmospheric pollutants. The development of regional pollution depends upon stable
air conditions, which prevent the pollutants from being dispersed
by wind, and the topography of an area, which may concentrate
the pollutants in a particular place. The key to regional atmospheric pollution is the total movement of a large volume of air.
Any aspect that reduces the dispersal of air, whether it is a slow-
moving body of stable air in a high-pressure cell or a topograph-
ical barrier, is important to understanding regional atmospheric
pollution.

Natural Air Pollution

Natural events in the environment, such as a volcanic eruption or a forest fire set by a lightning strike during a storm, may spew millions of tons of pollutants into the atmosphere. The pollutants may be limited to small areas or on rare occasions may be distributed worldwide.

Volcanic Eruptions

The more than 400 active volcanoes on the earth, as well as hot springs, fumaroles, geysers, and like features, constitute a practically continuous source of natural air pollution. From time to time the eruption of a volcano spews tremendous quantities of dust into the atmosphere to add to the particles already there.

Numerous examples can be cited of the effect of a single volcanic eruption on the climate of a region or even the world. In 1815 the Tambora eruption in Indonesia threw billions of cubic yards of dust over 15 miles into the atmosphere. At a distance of 300 miles, the pollution was so dense that it caused total darkness for three days. Because the dust penetrated the stratosphere, wind currents spread it throughout the world. For months, the dust in the upper atmosphere caused lengthy twilights and brilliant sunsets all over the world.

As a consequence of this volcanic activity, 1816 became known as "the year without a summer." In the middle latitudes of the Northern Hemisphere, temperatures averaged 0.6° C below normal; in parts of northern Europe, and particularly in the British Isles, the average temperature was 1.6°–3.2° C below normal. Cyclonic storm activity concentrated from eastern Canada across northern Europe, and from May to October 1816 there was almost continuous rain. In New England a heavy snow occurred between June 6 and June 11, and frost occurred every month of 1816. Crops that were not killed by the frost did not ripen or rotted in the fields, which led to widespread food shortages and food riots in Wales. By 1817 the excess dust had settled out of the atmosphere, and the climate had almost returned to normal conditions.

The greatest explosion in modern times was that of the Krakatao eruption in 1883, again in Indonesia. This massive explosion shot ash in the atmosphere at least 18 miles high, and it is estimated that about 13 cubic miles of material was ejected into the sky. A third of this dust fell within a radius of 30 miles, another third fell within 1,800 miles, and the remainder was spread worldwide. Even a year after the explosion much of the dust was still suspended at heights of about 9–10 miles, and it was three years before most of the dust settled to the level of the highest clouds. Although temperatures were lowered, a world system of distributing food prevented any severe shortage. In 1883 the *Scientific American* reported such unusual sky colorings as a green sun in India, a rainbow in a clear sky in the United States, and strikingly red sunsets in Peru.

Two of the most recent massive volcanic eruptions were Mount Saint Helens in Washington State in 1980 and El Chichon in Mexico in 1982. The ash from both eruptions, after being compacted by rainfall, had an estimated bulk of material equal to a football field piled 60 miles high. However, the type of eruption from the two volcanoes differed significantly so their effects on the atmosphere were strikingly different. The volume ejected at Mount Saint Helens was probably six to seven times greater than that at El Chichon, but the latter eruption had a far greater effect on atmospheric pollution because of three factors. The eruption of material was almost entirely upward at El Chichon, whereas at Mount Saint Helens the energy was expended horizontally more than vertically. At El Chichon the atmospheric conditions were more favorable to stratospheric penetration and the transport of the pollution, and finally, the El Chichon magma had an unusually high content of sulfur dioxide. At the National Oceanic and Atmospheric Administration's Mauna Loa Observatory in Hawaii, the laser radar indicated that the El Chichon cloud was more than 100 times denser than that from Mount Saint Helens. A satellite called the Solar Mesosphere Explorer indicated that the ash from El Chichon spread an uneven veil over much of the earth, extending from about 11 miles to 21 miles in altitude.

Scientists have raised the question, Has the ejection of volcanic ash into the atmosphere really affected the weather, or have the climatic changes been simply coincidental? Until the past century solar radiation could not be directly measured, so the effects of volcanic eruptions could not be verified. Since 1880, however, an instrument called the pyrheliometer has been used to measure the sun's radiant energy received on the earth. The results show marked fluctuations from year to year, and the periods when less energy reaches the earth coincide with the great volcanic eruptions. The marked decrease in temperature in

1884–1885 occurred after the 1883 eruption of Krakatao. Low temperatures in 1890–1891 were preceded by the eruptions of Bandai-san (Honshu, Japan) in 1888 and Vulcano (northeastern Sicily) in 1888–1890, those in 1902–1903 coincided with the eruptions of Mount Pelée and Santa Maria in the West Indies, and the 1912 low coincided with the eruption of Mount Katmai in Alaska.

Evidence from more recent eruptions confirms the earlier findings that volcanic dust in the upper atmosphere does decrease the direct solar radiation on the earth. This cooling effect occurs because the dust blocks out sunlight and has less ability to keep the earth's heat in the atmosphere than does carbon dioxide. Thus dust has a reverse greenhouse effect. Large particles of dust can hold in some heat, but particles of dust small enough to be carried into the stratosphere, and remain there for months or longer, cool the earth. Dust will cool the lower atmosphere rather than warm it as carbon dioxide does, because unlike carbon dioxide, dust actually changes the amount of energy the atmosphere receives and thus can lower the amount of radiant energy that the earth receives.

The greatest cooling of the atmosphere as a result of dust particles occurs in the middle and higher latitudes. The dust originating from volcanoes is generally carried poleward, over a period of weeks and months, by the flow of air from the tropics to the higher latitudes so there is an accumulation of dust in the middle and higher latitudes. Further, even if the particles were distributed evenly over the earth, the poleward areas would be more shaded than the tropical ones because sunlight has a nearly vertical path through the atmosphere in tropical areas but at the higher latitudes the sun's rays enter the atmosphere at a greater angle, and therefore have a longer path through the atmosphere. If the atmosphere is filled with dust particles, the dust particles are heated, which lowers the temperatures on the surface of the earth.

The ash from volcanoes can affect the climate of the world in other ways. For example, dust clouds can reduce the temperature in a particular region, and since prevailing winds carry the temperature between regions, the temperature of areas outside the direct influence of the volcanic dust cloud may also be affected. It has also been found that the dust particles can act as nuclei to form ice crystals in the upper atmosphere, increasing not only rainfall but the potential for snow.

Carbon dioxide is also ejected by volcanoes, and this gas may contribute to the greenhouse effect, which tends to raise temperatures. However, the amount of carbon dioxide produced by volcanoes is extremely small compared to the amount produced by the earth's biological processes. It therefore appears that such an effect of an eruption is limited to a few years unless a major eruption continues over a long period of time.

Smoke Pollution

Natural sources of smoke originate from the burning of grasslands and forests each year owing to lightning strikes and other causes. To illustrate, it has been estimated that in the African savanna, between the equator and the Tropic of Capricorn, more than 600 million tons of grass and low forests are burned each year during the dry period, and in the United States, the U.S. Forest Service estimates that an average of 120,000 forest fires occur annually, burning 600,000 acres of forest. The local and regional smoke pollution effects of such fires can be quite serious. A forest fire may last for many days and shroud the area downwind in smoke so dense that breathing is difficult. In the summer of 1987 a large number of isolated forest fires, most of which were set by lightning strikes, burned in the northwestern United States and covered the area with dense clouds. In the fall of the same year, from October 27 to about November 10, there were 9,000 forest fires reported in the 13 states, from Virginia to Texas, that make up the Forest Service's Southern Region, and the total acreage burned was equal to 500 square miles. In West Virginia, the smoke was so thick in some southern counties in early November that visibility fell below 100 feet. Smoke from fires in Kentucky and West Virginia was carried by wind currents into New England, and in Bridgeport and New Haven, Connecticut, about 500 miles from southern West Virginia, visibility was only about 1.5 miles. Radio and television stations advised people to stay indoors and use air conditioners to reduce the irritation caused by the smoke, and people with heart and lung conditions were urged to take extra caution.

The smoke from forest fires is limited to the lower layers of the troposphere so the first rains will cleanse the atmosphere and the wind will dissipate the smoke. Therefore, a region can have a high density of smoke pollution and within a short time be free of it.

Carbon Dioxide Eruptions

A number of lakes in the world contain huge volumes of carbon dioxide and other gases such as methane. Lake Kivu, perhaps the world's gassiest large lake, lies in the Great Rift valley of east Africa, between Rwanda and Zaire. The water of this lake, 60 miles long and 1,300 feet deep, is charged with more than 63 billion cubic meters of methane and five times as much carbon dioxide (CO_2). The methane gas in the lake is generated from bacteria in the sediments, and the CO_2 seeps slowly into the lake from nearby volcanic hot springs. These gases have been stable in the lake, but it is difficult to imagine the extent of the catastrophe that would occur if these gases were spewed into the atmosphere of the surrounding area. Plans are now being made to commercially exploit the methane but the technology must be carefully designed to avoid setting off a catastrophic gas explosion.

The devastating effect of a discharge of carbon dioxide from a lake was dramatically shown on August 21, 1986, when Lake Nyos in the west African country of Cameroon discharged some 1.3 billion cubic yards of CO_2 into the atmosphere. A water surge accompanying the gas burst gushed more than 250 feet into the air, and the lake was stained a deep red from an iron compound that was carried from the lake bottom by the escaping gas. The silent cloud of CO_2 moved ten miles down the valley; 1,700 people and hundreds of animals were killed, and 4,000-5,000 people left the area.

In an attempt to discover the cause of the disaster, isotopic analyses of the gas-charged water from Lake Nyos suggested that the carbon dioxide had slowly escaped from deep inside the earth and accumulated on the lake floor. The crater lake district of the country is a volcanically active region, and the last eruption had been in 1982 on Mount Cameroon. However, the experts could not agree on why the gas explosion had been so violent. One group of scientists believed that the carbon dioxide gas had accumulated in a volcanic pipe that connected the magma source with the crater above and that groundwater had come in contact with the hot magma to create a burst of steam known as a phreatic explosion, blasting a jet of gas and water upward through the lake. The majority of scientists, however, believed that the gas had accumulated slowly in the lake bottom and that a slight disturbance of the water, such as a rockfall, an earth tremor, a volcanic eruption, a strong wind, or even the seasonal cooling of the surface water, could have triggered the upward explosion of lake water.

A gaseous explosion such as that in Lake Nyos is rare but catastrophic when it does occur. Few lakes are gas laden, and in order to pose a threat, a lake must be deep and sheltered enough to stratify into distinct layers. Density differences, caused by chemicals in the water or by surface warming, can keep the layers from mixing, and dangerous gases can accumulate in the bottom layers. In many lakes such stratification is broken, especially in spring and fall when the cool waters sink and mix with the bottom waters, and in tropical areas strong winds or heavy rains will cause a lake's water to "turn over." The type of catastrophe created by the upheaval of the water in Lake Nyos can be averted by a proper monitoring of gaseous lakes, but this type of vigilance is lacking in a large number of Third World countries.

Air Pollutants Resulting from Human Activities

In the urban and industrial areas of the world, the atmosphere contains a wide variety of pollutants that are caused by human activities. Some of the chemicals are emitted directly into the atmosphere from identifiable sources, and others are formed indirectly through photochemical reactions in the air.

Direct Emission

Following is a list of direct pollutants and their sources:

Arsenic (As): from coal and oil furnaces

Benzine (C_6H_6): from refineries and motor vehicles

Cadmium (Cd): from coal and oil furnaces, burning of waste, and from smelters

Carbon dioxide (CO_2): from burning fossil fuels

Carbon monoxide (CO): from burning of coal and oil, smelters, and steel plants

Chlorine (Cl_2): from chemical industries; unites with hydrogen (H) to form hydrochloric acid (HCl)

Fluoride ion (F-): from steel smelters and plants

Formaldehyde (HCH): from exhaust pipes of motor vehicles and chemical plants

Hydrogen chloride (HCl): from incinerators

Hydrogen fluoride (HF): from smelters and fertilizer plants

Hydrogen sulfide (H_2S): from industrial plants such as refineries and pulp mills and sewage

Lead (Pb): from motor vehicles and smelters

Manganese (Mn): from steel and power plants

Mercury (Hg): from coal and oil furnaces and smelters

Nickel (Ni): from smelters and coal and oil furnaces

Nitric oxides (NO_X): from motor vehicles and coal and oil furnaces; changes to nitrogen dioxide (NO_2) quickly

Radioactive materials: includes a great nucleus of materials such as iodine-131, cesium-137, and strontium-90; major source is from accidents in the nuclear industry

Silicon tetrafluoride (SiF_4): from chemical plants

Sulfur dioxide (SO_2): from coal and oil furnaces and smelters

Photochemical Reactions

Following is a list of pollutants formed indirectly through photochemical reactions.

Hydroxyl radical (OH): formed in sunlight from hydrocarbons and nitric oxides; continued reaction with other gases produces acid

Nitric acid (HNO_3): formed in the atmosphere from NO_2

Nitrogen dioxide (NO_2): formed in sunlight from nitrogen oxide (NO) to produce ozone

Nitrous acid (HONO): formed by combining NO_2 with water (H_2O)

Ozone (O_3): formed in sunlight from nitric oxides and hydrocarbons

Peroxyacetyl nitrate (PAN): formed in sunlight from nitric oxides and hydrocarbons

Sulfuric acid (H_2SO_4): formed in sunlight from sulfur dioxide (SO_2) and hydroxyl ions (OH)

Major Air Pollution Incidents

Excessive air pollution can occur at all geographical levels from local to worldwide. Such incidents may be a response to an industrial accident, such as at Bhopal, India, or they may occur in response to a particularly stable atmospheric condition. The length of the incident will depend on a number of factors. A strong surface wind movement will disperse the pollutants relatively quickly; if the pollutants reach the stratosphere, they may be carried around the world before they disappear from the atmosphere.

Local

There have been many local air pollution situations that have been devastating. One of these occurred December 4–10, 1952, in London, England. On December 4 a high-pressure cell began to center on the city, shrouding it in several layers of clouds. From thousands of chimneys the unburned particles of coal floated into the atmosphere, and the smell of smoke was strong. In the following days the stagnant air did not carry the pollutants away, and smoke and moisture accumulated in the lower layers of the atmosphere. By December 6 people were experiencing discomfort in breathing because of the smog conditions. Dense fog blotted out the sky, visibility was only a few feet, and the humidity had risen to nearly 100 percent. As the smoke accumulated, coughing could be heard everywhere in the city. On December 7 conditions worsened. Patients with respiratory diseases crowded the London hospitals, and many died. On December 9 the high-pressure cell finally began to move, and with the wind blowing fairly steadily, the skies began to clear. By December 10 a cold front had passed over the area, bringing fresh, clean air from the North Atlantic. The emergency was over, but during those few days it is estimated that 4,000 Londoners died.

The United States experienced a major air pollution incident in Donora, Pennsylvania, on October 25–31, 1948. Donora, a heavy-industrial center, is located in the deep valley of the Monongahela River. During a period of extremely stable air, the industrial pollutants of sulfur dioxide and its oxidation products, along with particulate matter, accumulated in the atmosphere until a dense fog developed, which was not dissipated until it rained on October 31. Of a population of 14,000, more than 6,000 suffered some respiratory problems, 1,500 being seriously ill. Eighteen people died, all of them over 50 years of age, and 14 of them had had a previous history of respiratory illness.

In July 1976, when a stable air mass had developed over Milan, Italy, a chemical plant at nearby Seveso accidentally released a cloud of highly toxic dioxin (tetrachlorodibenzo-p-dioxin) into the atmosphere. The pollutant remained in the area for about three weeks, forcing the evacuation of 700 people, at least 500 of whom exhibited symptoms of poisoning. Pregnant women who were affected were advised to have abortions because the poison causes malformations in fetuses, about 600 animals were poisoned and had to be destroyed, and all contaminated crops had to be burned. Medical experts recommended that all residents of the area have periodic medical examinations for the rest of their lives.

Regional

On a number of occasions an excessive pollution concentration has covered nearly half the area of the United States. One such phenomenon, now known as Episode 104, occurred between August 22 and September 1, 1969. On August 22 the National Meteorological Center forecast a high air pollution potential (HAPP) for the upper Midwest. Such a condition occurs when the air near the earth's surface reverses its usual daytime decrease in temperature with altitude and for the first 1,200–1,300 feet or so the temperature increases with height, creating a temperature or thermal inversion. In this situation the cooler, pollutant-laden surface air is held near the ground by the warm air above it. If the winds during such an inversion are absent or very light, pollutants cannot move either upward or horizontally.

The high-pressure cell that started Episode 104 began to spread rapidly from its source area at the southern tip of Lake Michigan after August 22. On that day in Chicago, a Friday, the level of sulfur dioxide pollution was triple that of the previous day; by Saturday a massive high-pressure area extended south from the Great Lakes to the Gulf and eastward to the Atlantic. In this vast region, pollutants began to accumulate in the stable air from thousands of homes, factories, and businesses; thousands of miles of streets; countless sites of refuse burning; and motor vehicles. All cities of the region experienced problems of air pollution. On Thursday, August 28, air pollutants reached their highest levels, ranging from two to eight times greater than under normal conditions, and the number of respiratory ailments increased. On August 29 the high-pressure cell began to move, and by September 1 the sky was clear and the air pollutants had moved out over the Atlantic Ocean.

As the metropolitan areas—such as the vast megalopolis that extends from Boston to Washington, D.C., the southern Lake Michigan urbanized area centering on Chicago, and the metropolitan Los Angeles basin—have increased in size, incidents of regional air pollution have become more frequent. A haze of smoke and dust forms an umbrella over these regions. This shroud of pollutants, known as a "dust dome," is caused by a particular atmospheric circulation pattern that is caused by the marked temperature differences between urbanized and rural areas. The urban streets, parking lots, office buildings, factories, and homes of asphalt, brick, concrete, and steel radiate heat more readily than do the open fields and forests, and this urban-rural temperature differential is conducive to an atmospheric circulation pattern in which cool air slowly moves from the country into an urbanized area to replace the rising warm air. As a result, smoke, dust, nitric oxide, and other aerial "garbage" tend to concentrate in a vast pool above the urbanized area. The volume of pollutants may be 1,000 times greater immediately over an urban area than in the air of the surrounding countryside. When the air movement is slight, the pollutants appear as a stable mass of haze. If the wind speed is as great as eight miles per hour, the pollutants are pushed downward and horizontally into an elongated plume. Such a plume may extend for 150 miles or more.

Acid Precipitation

In the 1950s and 1960s studies conducted by scientists in the United States and western and northern Europe found evidence of acid rain in the physical environment. The pioneer endeavors of four scientists illustrate how sound scientific work established the foundation for the present-day study of the problems of acid precipitation.

In the early 1950s Eville Gorham, a Canadian, studied the ecosystem of the Lake District in northwestern England. One of his earliest discoveries was that the area was drenched with sea salt when the winds were from the Irish Sea to the west but that when the winds were from the industrial area to the south, the precipitation was loaded with sulfuric acid. This was an important observation, for while pH measurements of rain had been made in the late 1930s, it had not previously been realized that pollutants causing acid rain could be carried great distances. In 1955 Gorham published his finding that acid-forming pollution was carried great distances.

Gorham next wanted to study the ecological effect of acid rain, and because data were scarce, he decided to investigate the effects of sulfuric acid on mortality associated with respiratory diseases in British cities. In the late 1950s Gorham compared the incidence of bronchitis, pneumonia, and lung cancer in a particular English city with the data gathered by another researcher on the amount of air pollutants in the area. This study revealed that as acidic precipitation increased, there was a greater incidence of bronchitis. The greater the number of sulfate particles in the air, the more pneumonia, and the more tar, the greater incidence of lung cancer.

Gorham then investigated how the advent of acid rain might change life in an area. This study, conducted in Ontario, Canada, near Sudbury, was inconclusive and abandoned when Gorham moved to the University of Minnesota in 1962. Gorham believed that Minnesota was an ideal area in which to study acid rain because of its position between the industrial zone to the east and the cultivated prairies to the west, which could provide a source of alkaline dust that could neutralize the acidic conditions of the precipitation.

In order to investigate the problem, Gorham submitted proposals to the National Science Foundation, but they were rejected because the project was considered to be of little value. After months of delay the Energy Research and Development Administration approved funding. To implement the study Gorham chose three sites: one in the forested lake country of northern Minnesota, the second in central Minnesota between the forest and the prairie, and the third in western Minnesota in the open prairie and agricultural area. The study revealed that the acidity of rain and snow was least in western Minnesota—the area the furthest from the sources of pollution in the eastern industrialized areas. In this region the average precipitation had a pH of 4.66, an intermediate position between a pH of 4.5 and a pH of 4.8, the critical limits in an ecosystem. For Gorham, this study revealed that acid rain could develop from pollution sources hundreds, even thousands, of miles from the ultimate place of precipitation. Since the early 1960s Gorham has given hundreds of talks to alert the general public to the great potential problems of atmospheric pollution from the sulfur dioxide and nitric oxides and the resulting acid precipitation.

Svante Odén, a soil scientist at the Agricultural College near Uppsala in Sweden, has been identified as "the father of acid rain research." In the early 1960s Odén applied the knowledge of three related fields of science—limnology (the study of lakes), agriculture, and atmospheric chemistry—to the study of acid precipitation. He established a Scandinavian network for surface water chemistry in 1961, and after combining data from this network with data from the European Air Chemistry Network, Odén proposed a series of general trends and relationships concerning the distribution of acid rain. He published his findings in 1967 in Stockholm's prestigious newspaper *Dagens Nyhetu* and in 1968 in an ecology committee bulletin.

Odén's remark about an "insidious chemical warfare among the nations of Europe" received widespread attention in the press, which began the process of public education concerning acid deposition in Europe. Odén's analysis clearly demonstrated that, one, acid precipitation was a large-scale regional phenomenon in much of Europe with well-defined source and sink regions, two, both precipitation and surface water were becoming more acidic, and three, long-distance (100–1,200 miles) transport of both sulfur- and nitrogen-containing air pollutants was occurring in Europe. Odén also proposed a series of hypotheses about the probable ecological consequences of acid precipitation. These included:

1. Increased acidification of Scandinavian lakes and rivers

2. Decline of fish population

3. Acidification of soil systems

4. Displacement of nutrient cations from soils

5. Displacement of toxic metals from soils to surface water and groundwater

6. Acidification of groundwater

7. Decreased biological fixation of atmospheric nitrogen

8. Decreased growth of trees

9. Increased disease in plants

10. Accelerated corrosion and weathering of building and other materials

These ideas initiated the scientific study of the effects of acid rain.

In the early 1960s Gene E. Likens of Dartmouth College and F. Herbert Borman of Yale University were studying the life cycle of a small New Hampshire stream called Hubbard Brook. Their study of the ecosystem was comprehensive, and from the data collected it was evident that the rain falling in the region had made the stream acidic—but this was assumed to be a local phenomenon. However, after Likens visited Sweden in the late 1960s and discovered the work of Svante Odén and other researchers, he raised the question whether acid rain was of regional origin rather than local. In 1969 Likens moved to Cornell University and began making measurements in central New York State. The same kind of acid rain data was obtained as had been observed in New Hampshire.

In 1972 Likens, with Borman and Noye M. Johnson of Dartmouth, published a scientific paper presenting evidence for the first time that acid rain was prevalent over the entire northeastern United States. This paper initiated the investigation in North America as to the origin of acid rain. The power utilities made a great effort to refute the work of Likens and others, and in the mid-1970s two papers published in the highly respected scientific journal *Science* said that sulfuric and nitric acids were not the principal acids present in acid rain and therefore power plants and other major users of fossil fuels could not be responsible. Likens and other scientists have rebuffed those conclusions, and it is now almost universally accepted in the scientific community that acid rain is caused by sulfur and nitrogen emissions from the burning of fossil fuels.

Sulfur Dioxide (SO_2) and Nitric Oxides (NO_X)

Sources of Sulfur Dioxide

All plants and animals contain some sulfur, and fossil fuels are formed when dead animals and/or plants are converted into liquids (oils), gases (natural gas), or solids (peat and coal). When fossil fuels are burned, the long-stored sulfur is released as sulfur dioxide (SO_2). Many mineral ores also have a small percentage of sulfur, and when these ores are smelted, SO_2 is emitted. The sulfur content varies among the fossil fuels and ores. Crude oil may have from 0.1 to 3.0 percent sulfur content, and the sulfur in coal and ores varies from about 0.4 to 5 percent.

According to the U.S. National Research Council, about one-half the SO_2 in the atmosphere comes from natural sources. But it is estimated that 100 million tons of SO_2 are emitted worldwide every year from coal- and oil-fired power stations, industries that consume fossil fuels, and smelters. The geographical concentration of SO_2 emissions varies, with the highest levels originating in the industrial centers of Europe, North America, and the Far East.

Within each of these continental areas there are major source areas. In the eastern half of the United States the leading states in SO_2 emission are Ohio, Indiana, and Illinois. In these three states there are thousands of points of emission, but the 20 largest coal-fired power plants emit 20–25 percent of all SO_2 in the eastern United States. In the Ohio Valley two of the largest plants, the Kyger Creek and the James M. Gavin, emit 570,000 tons of sulfur dioxide annually from their tall smokestacks.

In Canada, smelters are the largest emitters of SO_2, accounting for 45 percent of the nation's total. The largest single source of SO_2 produced by humans in the world is the copper and nickel smelter complex at Sudbury, Ontario. It is estimated that the Sudbury smelter emitted more SO_2 between 1969 and 1979 than was discharged from all the volcanoes in the history of the earth. During this period the greater-than-mile-high superstack spewed 5,000 tons of SO_2 each day into the atmosphere. Since controls were introduced, the emission has been reduced to less than 1,950 tons per day.

Sources of Nitric Oxides

More than 90 percent of the nitric oxides (NO_X) emitted into the atmosphere comes from human sources, and the single major source is the combustion of fossil fuels. In the United States about 15.5 million tons of NO_X are emitted annually into the atmosphere.

Coal contains about 1 percent nitrogen by weight. When high-temperature combustion occurs, the nitrogen combines with oxygen to form NO_X. About 55 percent of NO_X emissions comes from power plants and industrial and residential users of coal. The remaining 45 percent comes from mobile sources such as motor vehicles, planes, and trains. Of the amount emitted, about 6.6 million tons are deposited in North America as acid nitrates in precipitation; another 4.4 million–8.8 million tons are dry deposited. Each year millions of tons are deposited in the oceans.

Conversion of Sulfur Dioxide and Nitric Oxides

Once SO_2 and NO_X are emitted into the atmosphere, complex chemical processes occur that transform them into gaseous or liquid forms of sulfuric acid (H_2SO_4) and nitric acid (HNO_3). The amount of acid produced depends on weather conditions such as humidity, cloud covering, and sunlight and on the presence of other pollutants.

There are two phases involved in the formation of acids—dry and wet—and in both phases SO_2 and NO_X are converted to sulphate and nitrate. In general, the dry, or gas, phase occurs relatively close to the point of emission. The wet, or aqueous, phase involves reaction with water droplets and can occur hundreds or even thousands of miles from the point of emission.

In the gas phase oxides are produced in several ways, usually with the aid of a photochemically generated radical. (A radical is an atom or group of atoms containing one or more unpaired electrons that exist for a short time before they combine to produce a stable molecule.) The hydroxyl radical (OH) is the principal agent to combine with both SO_2 and NO_2 to produce sulfuric (H_2SO_4) and nitric (HNO_3) acids. The OH radical is formed under the influence of sunlight—the rate of conversion declines rapidly after sunset—in several ways. The most common way is a complex reaction involving ozone. The oxides are broken down by sunlight, forming a highly reactive oxygen atom that combines with the diatomic oxygen (O_2) molecule to form ozone (O_3). Ozone is an unstable molecule; when it breaks down, oxygen atoms are released, and they may react with water to produce the OH.

The rate at which the acids are formed is thought to depend upon the concentrations of OH present, but direct measurement of the OH in the atmosphere is difficult because of the instability of the radical. The presence of water is essential, and the oxidation on a cloudy, rainy day is far more rapid than on a clear day. The presence of a trace metal catalyst or strong oxidizing agents is also vital. Manganese appears to be the only metal that can act as a catalyst in cloud droplets. Of the oxidizing agents, ozone and/or hydrogen peroxide (H_2O_2) appear to increase the conversion to acids. Although this process has been intensively studied, questions still remain.

Atmospheric Transport

The atmospheric movement of SO_2 and NO_X pollutants is a complex and politically charged issue, and scientific studies since the early 1950s have revealed a number of findings. Local emissions of pollutants will cause some acid rain to fall in that area, but in addition, the pollutants in many areas are transported hundreds, or even thousands, of miles downwind, across national and international borders.

The distance pollutants are transported depends upon a number of factors: the wind speed, weather conditions, the chemical state of the pollutants, and the height of the smokestacks. Obviously a high wind can transport pollutants hundreds of miles in a few days, but weather conditions are also a significant factor. If the skies are clear the possibilities of long-distance transport of pollutants is enhanced; in contrast, if there is heavy precipitation, the pollutants are quickly removed from the atmosphere. The chemical state of the pollutants also affects the distance they are transported. Acid sulfate is transported further than SO_2 because it is less strongly absorbed by the ground, so atmospheric conditions that favor the formation of acid sulfate will increase the transport distance.

One of the most important factors in the atmospheric transport of SO_2 and NO_X is the height at which they are emitted into the atmosphere. If the pollutants are emitted into the "mixing layer" of the atmosphere—that is, the layer closest to the ground where there is good vertical mixing—the pollutants will not be transported great distances. The mixing layer is typically below 3,000 feet and is visible as a blanket of polluted air covering an urban area. Pollutants in this lower layer make contact with the ground relatively soon and are generally transported a distance that is roughly proportional to the height of the smokestack.

Pollutants that are emitted above the mixing layer are effectively removed from ground contact and can be transported great distances by the strong winds aloft. Since the passage of the Clean Air Act of 1970 in the United States, the Environmental Protection Agency (EPA) has continued to measure power plant compliance with the law by measuring the amounts of SO_2 and NO_X a plant emits locally, but the height of smokestacks has also been increased greatly, with a few reaching over 1,200 feet. As a result, the local effect of emissions is reduced, but the possibilities of long-distance transport of the pollutants is greatly increased.

Types of Acid Deposition

Acid rain was originally thought of in terms of the acidity of precipitation, but this is too limited a viewpoint. Emissions from industrial and transportation sources are mainly gaseous, and they can be transported great distances and returned to the surface without being part of a cloud covering. Also, the gases may undergo chemical and/or physical transformation into finely divided liquids or solids, called aerosols, and these, in turn, may be intercepted by features on the earth's surface. Thus acid rain encompasses wet deposition, mainly through rain and snow, but can also include the interception of fog by buildings, trees, and landforms and dry deposition.

The quantity of wet deposition has been measured with varying degrees of reliability for decades. In contrast, dry deposition has been measured only sporadically, but estimates indicate that over the eastern part of North America, the totals of wet and dry depositions are approximately equal in magnitude. Dry deposition is, in general, greater than wet near the emission sources. An additional complexity for dry deposition is that the chemical species behave differently. In general, gaseous nitric and sulfuric acids and ammonia are removed more efficiently than their particulate forms, nitrates and sulfates.

Acidity of Acid Rainfall

There is now a nearly worldwide system for the collection and measurement of acid precipitation, and in the United States and Canada, there are over 140 collectors operated by government, university, and private research institutions. The acidity or alkalinity of precipitation is measured according to the pH scale, the commonly used yardstick for such measurements.

On the pH scale, the number 7 indicates an absolutely neutral substance—one that is neither acidic nor alkaline. A weak alkaline substance, such as baking soda, measures about 8 on the scale, and a strongly alkaline substance, such as household ammonia, measures 12. Numbers lower than pH 7 indicate acidic substances. A mild acid, such as tomato juice, measures 4.3; a somewhat stronger acid, such as household vinegar, about 2.8; and a strong battery acid, close to pH 1. It must be remembered that the pH scale is logarithmic. Each whole number increase or decrease on the scale indicates a ten-fold increase or decrease in acidity or alkalinity. Thus, a pH 4 solution is 10 times more acidic than a pH 5 one and 100 times more acidic than a pH 6 solution.

Most natural rainfall is not neutral but slightly acidic because rain combines with carbon dioxide to produce a weak carbonic acid. As a consequence, normal rain tends to measure between 5.6 and 5.7 on the pH scale. It is also possible for rain to be slightly alkaline if there are limestone particles in the atmosphere. Under these conditions the pH rating may be about 7.1.

Today the precipitation falling over eastern North America and western Europe, as well as other industrial areas, ranges from 10 to 100 times more acidic than normal precipitation. According to the EPA, the acidity of rainfall over eastern North America now averages about 4.5 on the pH scale. For particular precipitation periods, the pH rating will vary tremendously. In the White Mountains of New Hampshire, meteorologists have reported that at least once every year samples of rain have had an acidity of pH 2.8, as acidic as vinegar. On April 10, 1974, a rainstorm over Pitlochry, Scotland, tested at a pH of 2.4, and values of 2.7 were reported in Norway during the same month. The lowest recording in a single rainstorm occurred in the fall of 1978 in Wheeling, West Virginia, when pH values of less than 2 were recorded during a three-day drizzle. This rain was thus 5,000 times more acidic than normal.

Although the data are still meager, there is increasing evidence that the acidity of precipitation is increasing steadily in the United States, Europe, and other industrial regions, although the effects of this acidity vary from region to region. In areas in which there are alkaline minerals, the effects of the acid precipitation are buffered, and little or no damage will be done. In contrast, in areas that have little or no neutralizing capacity, the effects of the acidification are great.

Environmental Effects of Acid Precipitation

Water is one of the most fundamental resources on the earth. All living things require water to exist. When such a resource is endangered by pollution, there is an absolute need to determine the changes that have taken place and to develop remedies to control future deterioration. Acid rain is an intricate scientific puzzle that requires the entire range of human endeavors to control its effects on the total environment. The major adverse effects that have been measured thus far include the acidification of streams and lakes, damage to the forest environment, degradation of the soils, damage to man-made materials, adverse effects on human health, and the leaching of toxic metals into the environment. It is difficult to predict the future effects of acid precipitation on something as complex as the total ecosystem, but the potential for great ecological devastation exists.

Streams and Lakes

In the past several decades acid rain in eastern North America and western and northern Europe has caused the acidification of sensitive streams and lakes, resulting in changes in the aquatic ecosystem and the decline and death of fish. Normal lake water will have a pH value of around 5.6 and above. When the measurement reaches 5, a lake is considered to be acidified, and when the level reaches pH 4.5 or below, the lake is usually considered dead.

Most rain in the eastern half of the United States is more acidic than pH 4.5, frequently measuring 4.3–4.0. Under such conditions why are only some of the lakes acidified and others not? The reason is that in large areas there is alkaline material that neutralizes the acidity, the most common alkaline rocks being limestone and dolomite. However, there are also vast areas in North America and Europe in which such rocks as sandstones, shales, granites, gneisses, and quartzite exist, and they are poor buffers. These areas include the vast Canadian and Scandinavian shields, the Adirondacks, parts of the Appalachians, and much of the mountainous regions of western North America and central Europe.

EASTERN CANADA   Acid precipitation has now been reported in scientific studies as occurring from Alberta eastward to Newfoundland. In Ontario the Ministry of the Environment has reported that 1,200 lakes have lost their fish population because of acidification, about 3,400 more are approaching that state, and an additional 11,400 are considered at risk. It is predicted that by the year 2000 an additional 48,500 lakes in the province will have lost their fish population. The Muskoka-Haliburton region, a leading tourist area, is among the areas most threatened. The mean pH of rainfall in this region now ranges from 3.9 to 4.4.

To test the effect of the acidification, the Canadian government has conducted a controlled study of a lake in western Ontario since 1974. The isolated lake is called simply Lake 223. At the beginning of the study Lake 223's water had a pH value of 6.6.

Each year, government scientists add measured doses of sulfuric acid to the lake while simultaneously calculating any further additives of acid from precipitation. In the first year the artificial addition of acid to the lake simply speeded up the removal of what little alkaline material still existed in the lake. Since then, the amount of acid placed in the lake has been the equivalent of the amount falling in precipitation in eastern Canada. As a result the pH has been lowered by about .25 units per year, more than doubling the acidity.

As the lake became more acidic its aquatic life was greatly altered. At pH 6 the tadpole shrimp began to experience stress. Normally, maturing shrimp undergo a series of molts of their shells, but that rate began to slow, and many died before reaching maturity. When the pH reached 5.8 the shrimp and fathead minnows—both primary food for the native lake trout—vanished.

There has been one unexpected change in the ecosystem. As the pH decreased from 6.6 to 5.05 there was a slight increase in algal plants and number of zooplankton. Scientists believe that this increase was the result of the increased acidity removing natural humic materials that lightly stain the water. The sunlight was then able to penetrate more deeply into the water to stimulate this type of growth. It is thought that as the acidity rises, these new tiny plants and animals will be killed.

The stocking of acidified lakes has had disastrous results. To illustrate, in the spring of 1966 more than 4,000 young pink salmon were stocked in Lumsden Lake, a small lake in the center of the Killarny Park wilderness of Ontario. An isolated lake, its clear, 60-foot-deep water made it physically perfect for salmon and other sporting fish such as trout, perch, and lake herring. To guarantee that the fish would not migrate out of the lake, a wire screen was placed across the outlet. At the time the salmon were stocked some white suckers, potential food for the salmon, were taken from the lake. The suckers were abnormally small and appeared unsuccessful at breeding. Although this problem was investigated, no real answer to their smallness was found. In the spring of 1967 a check of the salmon population revealed that not a single fish was living.

In 1968 the experiment continued with the tagging of 100 suckers in Lumsden Lake and 60 more in neighboring lakes. One year later, in 1969, only a few of the tagged fish could be found, and in Lumsden Lake all one-year-old suckers had disappeared. Between 1969 and 1971 the fish population of five tributary lakes was investigated, and by 1971 all fish had disappeared. A study of the pH of Lumsden Lake revealed that between 1961 and 1971 the pH level had fallen from 6.8 to 4.4—a highly abnormal 100-fold increase in acidity in one decade.

The decline of the fish population in Lumsden Lake and most of the lakes in the Killarny Park wilderness basically indicates the way fish and lakes die from acidification everywhere. The process and pattern vary somewhat, but the end result is, first, a decline in total fish population; second, the disappearance of a particular species; and eventually, the loss of all fish.

There is great variation in the acidification of the rivers in eastern Canada. A major question has concerned the effect acid precipitation has had on the 60 Atlantic salmon rivers on the north shore of the St. Lawrence in Quebec. In 1981 Canada's Department of Fisheries and Oceans began to study four of these streams—the Petit Saguenay, the St. Marguerite, the Escoumins, and the Petit Escoumins. The study revealed that three of the streams had not been affected by acid precipitation but that the fourth, the St. Marguerite, had high levels of aluminum although the pH level varied from 6.4 to 6.7.

In contrast to the Quebec rivers, the Nova Scotia Department of Fisheries and Oceans found that ten Atlantic salmon rivers in that province were acidified and unable to support fish. The pH level of all of these rivers was below pH 5, and some were below 4.7. These rivers had been famous for salmon for hundreds of years, and there had been fairly steady catches until the 1950s, but all fish had disappeared by 1970. An electrofishing survey in 1980 failed to reveal any sign of Atlantic salmon reproduction, and these salmon rivers are now considered dead.

In another 11 Nova Scotia rivers, in which the pH ranges from 5 to 4.7, Atlantic salmon still maintain self-sustaining runs, but their number has decreased. In addition, there are 9 other rivers in Nova Scotia with pH values of 5.1–5.4 in which the salmon are considered threatened.

EASTERN UNITED STATES   Studies of the acidification of streams and lakes in many parts of the eastern half of the United States show a situation similar to that in eastern Canada. One of the major areas of acidification is the Adirondack Mountains in New York State where the New York Department of Environmental Conservation (NYDEC) has documented 212 lakes and ponds that are incapable of supporting fish life because of acidification.

In addition, another 250 lakes and ponds have been judged to be in danger of losing their fish. In 1981 the NYDEC reported that 40 percent of the 114 streams sampled after the 1980 snowmelt showed "critical" or "endangered" levels of acidity. Fifteen percent of the streams had pH levels below 5 and were devoid of fish. An additional 15 percent were termed endangered because they had a pH of 6 or less, and most of those streams had no fish. Most of the streams and lakes with the highest acidity were found in the southwest closest to the winds from the industrial interior. Other sensitive areas in New York State include the Catskill Mountains, Tug Hill Plateau, the Shawangunk Mountains, the Hudson Highlands, and Long Island.

In Massachusetts, streams supporting sport fishing and a number of reservoirs providing drinking water have problems of acidification. In a 1980 state survey of 154 reservoirs, more than 60 percent proved vulnerable to acid precipitation. A main concern is Quabbin Reservoir, which is the source of water for the 2 million people who live in the Boston metropolitan area and in addition supports a $1-million-annual sports fishing industry. The pH level of the reservoir's surface water often drops below 5. Total alkalinity is very low, ranging from one to four parts per million, and there is concern that the acidic drinking water could corrode pipes. Several hundred thousands of dollars are spent each year to raise the alkalinity of the reservoir, but given the present trends of acid precipitation, Quabbin will lose its fish population sometime in the late 1990s.

Plymouth County and Cape Cod in the southeastern part of the state illustrate the variations that occur in water problems. Some ponds and lakes show no sign of acidification while others are acidified. The situation depends on the local geology. Because of the low alkalinity of only 8–14 parts per million in certain ponds, liming has long been practiced to reduce the acidity. There is, however, at least one pond, Hathaway Pond in Barnstable, that is so highly acidified that it is no longer limed or stocked with trout. It is a dead pond.

The acidification of the lakes and streams of Vermont is similar. A 1979–1980 survey of the 120 lakes in the state revealed that the majority of them were potentially susceptible to acidification because of low alkalinity. In 1979 the natural water consumed by the town of Bennington tested below the federal standard for drinking water of pH 6. As a result, the town of Bennington had to spend $37,000 to treat the water with sodium bicarbonate and sodium hydroxide to reduce its corrosiveness.

Acid precipitation is not confined to the traditional industrial states but extends southward to Florida and westward to Minnesota. In the Great Smoky Mountains National Park, the beautiful blue haze that is caused by the lacquers and oils emitted from the forest are now rarely seen. Instead, a great gray-brown haze, composed of industrial chemicals and particulates, covers the region. The southern Appalachian region now has the second highest frequency of air stagnation in the United States, second only to the Los Angeles basin, and the annual pH of precipitation in the Great Smoky Mountains National Park decreased from 5.5 in 1955 to 4.4 in 1973 to 3.7 in 1980.

The ecological systems of the park are therefore threatened. Brook trout, the only native salmonid, is endangered, and most of the rainbow trout contain high concentrations of mercury. In April 1982 a sudden surge of acidic stream water coming from the park killed 1,000 rainbow trout in a hatchery on the Cherokee Indian reservation, and in lakes outside the park, smallmouth bass have abnormal backbones, a condition normally associated with aluminum toxicity.

It is estimated that there are 16,000 lakes in Michigan that are susceptible to acid precipitation. More than half of the water bodies on the Upper Peninsula have an alkalinity of only about ten parts per million, and scientists now believe that the greatest impact on these lakes and ponds comes in the spring when the heavy snow accumulations melt in a short period of time. It was once thought that the snows were clean, originating from storms out of the southwest, but it is now known that the winter storms often come from the industrialized areas along the southern Great Lakes. Snowfalls with a pH value as low as 3.9 have been recorded, and the medium pH for these snowfalls is about 4.5.

WESTERN UNITED STATES   Acidification of lakes and streams is now also found throughout the western half of the United States. In Colorado a number of studies have revealed widespread acidification. In a small stream west of Denver, Como Creek, at an elevation of 9,000 feet, the pH dropped from 5.4 to 4.7 between 1974 and 1978. It was once thought that the acidification was due to polluted air moving up slope from the Denver area, but more recent studies have revealed that lakes in western Colorado also have very high levels of acid. It is now believed that the acid precipitation over Colorado is caused by emissions from coal-fired plants, not only in Colorado, but in surrounding states. The lakes of the Rocky Mountains are poorly buffered because of their granitic bedrock so their acidification advances in much the same fashion as in the Adirondacks and on the Canadian Shield.

On the west coast, too, there is acidification of the lakes. In a study of the Cascade and Olympic mountains in Washington State, 24 of the 68 lakes sampled had a pH level lower than 6. In addition, seven lakes in the Alpine Lakes Wilderness in the Cascades east of Seattle have acidity levels below 5.5. In California a number of lakes in the Sierra Nevada have pH levels of 5.8 and high levels of aluminum, and additional studies are needed to document the effects of acidification in this ecosystem.

NORTHERN EUROPE   In a study that lasted from 1972 to 1980, Norwegian scientists sought to show precisely the effect of acid precipitation on freshwater fish and forests. In 1980 the study concluded that the decline of the fish population was of major importance to the people of the region. In many areas of southern Norway the fish population had declined by one-half, and in other areas fish almost completely disappeared between 1940 and 1980. If acidification continues the fish population in hundreds of additional lakes will be threatened. The decline of fish is related to a number of factors such as high egg and fry mortality and physiological stress resulting from toxic combinations of water acidity and aqueous aluminum.

Other European studies have revealed considerable information concerning the effects of acidic levels on different species of fish. At pH 5.5 brook, brown, and rainbow trout experience significant reductions in egg hatchability and growth; at pH 5.4 largemouth bass, smallmouth bass, walleyes, and rainbow trout disappear from streams and lakes, and salmon and trout have reduced populations. Below pH 4.8 lakes and streams will not support a fish population. In Norway studies have revealed that Atlantic salmon is the most sensitive, followed in order by sea trout, brown trout, perch, char, brook trout, pike, and eels.

Fish die not only because of acidification of their waters, which interferes with the salt balance of their body tissue and blood plasma, but also from toxic metals. Acid precipitation acts as a mobilizer of aluminum, manganese, zinc, nickel, lead, mercury, and cadmium, and the mobilization of these metals provides poisons that destroy fish systems. In Sweden the government has cautioned the population not to eat fish contaminated with toxic metals.

In addition to losing fish, an acidified body of water also loses many other organisms including certain types of algae, crustacea, mollusks, and insects, many of which are in the fish food chain. Each species is affected by a different pH level; for example, the stream stoneflies and magflies generally disappear at pH 5.

An acidified lake is often a beautiful crystalline blue. Many such lakes are carpeted with mats of algae, which can survive in acidified water, and beneath this algae bacteria that can live with oxygen thrive, producing gases that bubble up to the surface. Thus, leaves that fall into acidified lakes are preserved because the bacteria and fungi that normally break down the leaves are not present. As a result a leaf mulch builds up, even to the point of choking the body of water.

CAN ACIDIFIED LAKES BE RESTORED?   Because the acidification of lakes results in the loss of important commercial and recreational fisheries, fishing scientists and recreational managers have become increasingly interested in whether or not acidic lakes can be restored, and there have been attempts to neutralize acidic lakes and streams in Sweden, Norway, Canada, and the United States. In Scandinavia projects were begun as early as the 1920s to reduce the mortality in the salmon hatcheries, and in the United States liming has been practiced on a limited scale since the 1940s to increase fish production in acidic lakes.

The most commonly used neutralizing agent is some form of limestone ($CaCO_3$ or $CaOH_2$), hence the term *liming* is applied to this process. There are a number of ways to treat an acidified lake. The most common method is to apply the neutralizing agent directly on the surface of the water. Other methods include injecting the agent into the lake sediment, applying it to the watershed surrounding the lake, or placing the neutralizing material in biologically critical areas. Boats are commonly used to apply the neutralizing agent, but in isolated areas airplanes and helicopters are employed.

There are several advantages of using the liming process. Limestone is a relatively inexpensive material and is widely distributed, the technique is simple, there are reasonably accurate methods for determining the dose required to meet specific water quality targets, and the response of the system to treatment is rapid.

Although liming is several decades old, research on the effects of the technique has only recently been initiated. The first research program was established in Canada in 1973 by the Ontario Ministry of the Environment. Two small lakes, Middle and Lohi, were selected because they were highly acidified, including having high concentrations of toxic metals. They were surrounded by exposed bedrock but had limited human access even though they were within ten miles of Sudbury. In 1973 and 1974 pure lime and limestone were spread on the surface of the lakes' waters. By the end of 1974, 38 tons of the neutralizing agent had been placed in the water of the lakes. By 1975 the acidity of Middle Lake had been reduced a 100-fold, and the concentrations of toxic nickel, copper, and zinc had been lowered as the limestone removed the metals from the water and placed them in the sediment on the lake bottom. Results in Lohi Lake were not as satisfactory, and liming of this lake continued in the summer of 1975. The researchers added fertilizer to increase the growth of organisms that could provide a food source for the fish.

By August 1976 the researchers felt that Middle Lake was sufficiently neutral so that restocking of fish could begin, and 2,500 small bass were placed in the lake. But in the spring of 1977 not a single fish was alive. Lohi Lake, after being limed up to a near-neutral condition, lost 1,200 brook trout in the four summer months of 1977. Although the results of the restocking were disastrous, the scientists decided to continue the experiment. To provide physical and biological conditions that would be as close to nature as possible, the restocked fish were first placed in an enclosed pond in a nearby neutral lake. The trout were well fed, and none died in the holding area. Half the fish were then trucked to Middle and Lohi lakes, and the other half were taken to neutral lakes. The fish were slowly lowered into the lakes in cages in order to acclimatize them. Scuba divers monitored the fish, but within 24 hours, the trout placed in Middle and Lohi lakes were swimming in confusion, and the first deaths occurred within 48 hours. At the neutral lakes, the divers monitored the fish for the entire summer, and none of the new fish died.

In Middle and Lohi lakes the fish died of copper, zinc, and nickel poisoning because the liming had failed to reduce the toxic metal concentrations in the sediments. The continuing acid rain brought new acidity, and more toxic metals also entered into the lakes from their watersheds. The conclusion was that liming was completely ineffective and that as long as acid rain continued, the lakes would remain dead.

In the United States, government agencies in eight states have initiated approximately 100 very small liming operations since 1970. New York State leads in the number of projects with about 70 lakes in the Adirondacks being treated to neutralize high acid conditions. Funds have been limited for all these projects, and most of them have not had a highly scientific approach. There were only superficial analyses of the conditions before liming began, and there has been little post-treatment monitoring. None of these projects has added greatly to an understanding as to whether acidified lakes can be restored.

The most important work in attempting the restoration of acidified lakes was started by the Electric Power Research Institute. The research project, known as the Lake Acidification Mitigation Project (LAMP), involves three clear-water Adirondack mountain lakes. Two of the lakes, Cranberry Pond and Woods Lake, are highly acidic and have no fish population. The third lake, Little Simon Pond, is moderately acidified and supports some brook and lake trout. This private industry effort is being supplemented by the U.S. Fish and Wildlife Service, which is sponsoring a ten-lake study in the same region. The objectives of these studies are to examine the fish populations' response to liming and the effectiveness of the process in reducing the acidity of the lakes, evaluate the effectiveness of restocking, and provide management guidelines for lake management. The initial efforts are being expanded with the implementation of a national Acid Precipitation Mitigation Program (APMP), which is being coordinated by the Eastern Energy Land Use Team of the U.S. Fish and Wildlife Service. All of these programs are too new to evaluate their effectiveness.

In Europe, Sweden and Norway have also been conducting research on chemical neutralization techniques. In 1977 Sweden initiated a state-funded National Swedish Liming Project and then conducted over 450 liming projects in the treatment of 1,500 bodies of water. Most of these projects were operational rather than scientific, so few data were obtained concerning the effectiveness of the liming program. In 1982 the Swedish government initiated a larger project with the goal of treating 20,000 lakes. The Norwegian government has also sponsored field and laboratory experiments, almost all directed toward protecting the brown trout and Atlantic salmon populations.

Conclusive evidence has not been obtained that liming is a permanent solution to the problem of acidified lakes. Some temporary results are encouraging, but local conditions are important, and each lake must be analyzed to find an effective treatment. When toxic metals are present as well as acidification, the programs have not been successful. Costs are also increasing and now total between $500 and $1,000 per acre of surface water for a single season of treatment. It now appears that as long as acid rain continues to fall in a region, the restoration of lakes and streams remains in jeopardy.

Forests

Since early 1970 there has been widespread evidence of a decline in the growth of evergreen forests in the eastern part of the United States and western Europe. In Europe alone, 8.5 million–10 million acres of forests show signs of deterioration.

GEOGRAPHICAL DISTRIBUTION Because of the widespread decline in forest growth, a number of countries have attempted to document the destruction. Of the scientific studies, those in West Germany have been the most thorough. In 1982 that country's federal minister of food, agriculture, and forestry estimated forest damage at 1,300,000 acres, about 8 percent of the country's forests. In 1983 a new study found that 6,250,000 forest acres were affected, 34 percent of the total. This increase is no doubt partially owing to more thorough investigation.

The most affected areas of West Germany are in the heavily wooded regions of Bavaria and Baden-Württemberg. Nationwide, three-quarters of the fir trees are affected, and damage to spruce and pine rose from 9 to 41 percent in less than five years. These three species compose two-thirds of West Germany's forests. Since most of the affected trees are immediately removed from the forests, the devastation is striking.

Many other countries of Europe also reported forest damage. In Czechoslovakia some 500,000 acres are damaged, and in the Erz Mountains 100,000 acres are reported killed. In the Krokonose National Park near Prague 85,000 acres of spruce are dying, and the species has stopped regenerating in the mineral soils. In Poland 1,200,000 acres are affected. Forest researchers in Katowice have reported that fir trees are dead or dying on 450,000 acres, and the spruce forests around Rybnik and Czestochowa, also in the southern industrial region, are completely gone. Environmentalists have warned Poland that as many as 7,500,000 forest acres may be destroyed by 1990 if the present industrialization plans, based on the burning of high-sulfur brown coal, are continued.

In 1983 acute deterioration of the forests of the Netherlands was reported, with pine and fir in the eastern part of the country being heavily damaged. In Switzerland about 25 percent of the fir trees and 10 percent of the spruce died between 1983 and 1984. In East Germany some 12 percent of the forests are believed to be affected, and forest damage has also been reported in France, Italy, the United Kingdom, Austria, Yugoslavia, and Romania.

By 1983 early signs of forest damage had begun to appear in northern Europe. In Sweden spruce and pine appear to be the most severely damaged, and although surveys are not complete, it is believed that about 10 percent of the forest is affected. There are reports that spruce is damaged in southern Norway.

In a rare environmental report from the Soviet Union, the Communist paper *Pravda* reported in 1983 that vast forests were dying because of air pollution in the region around the motor-vehicle manufacturing center of Togliatti, 800 miles east of Moscow. According to the report, forests along the Volga River may soon disappear.

In the United States forest damage is most evident in the Appalachian mountain ranges in the east and in the Sierra Nevada in California. Scientific studies have documented not only tree diseases and destruction of the forest but also sustained declines in growth. Damage is most severe in the high-elevation forests of New York, Vermont, and New Hampshire. Of these three states, the greatest documentation of forest decline is on Camel's Hump in the Green Mountains of Vermont. In studies conducted between 1965 and 1979 researchers found that seedling production and tree density had declined by about half; in 1979 over half of the spruce trees on Camel's Hump were dead.

CAUSES OF FOREST DAMAGE   The visible damage to an evergreen first takes the form of yellowing and an early loss of needles. These indicators are followed by deformed shoots, a deterioration of the roots, a progressive thinning of the tree crowns, and ultimately, the death of the tree. Because the massive deterioration cannot be explained by natural events, many scientists have investigated the role of air pollutants, and of the possibilities, acid rain has provided strong circumstantial evidence. In West Germany the damage to the forests is greatest on the west-facing mountain slopes which are exposed to the most acid rain and fog. Yet trees are suffering in other areas, too, even though the concentration of acid rain is low.

In other areas scientists have firmly documented that tree disease is linked to the availability of ozone and other pollutants in the family of photochemical oxidants. Ozone forms when nitric oxides react with hydrocarbons in the presence of sunlight, and a long-time study of the effect of ozone on the pine forests in the San Bernardino Mountains east of Los Angeles revealed that the yellow-brown photochemical smog had a devastating effect on the trees. A decline of the forests was evident by 1950, and since then losses of the Ponderosa and Jeffrey pines have been dramatic. Losses have been greatest on the westward-facing slopes, which receive the highest pollutant doses. Between 1941 and 1971, when compared to the 1910–1940 period, the radial growth decline was 38 percent, and in the areas receiving the largest ozone doses, the marketable value of 30-year-old pines declined by 83 percent.

In spite of the evidence that air pollutants are causing forest damage, the evidence is not conclusive, and skeptics point out that the decline of certain species occurred before air pollution was evident. To illustrate, in the 1930s and 1940s a dieback of birch occurred in Nova Scotia, Maine, and the Adirondacks, but the birch has recovered and is flourishing again. There have also been periods when the red oak and maple declined. Presently the mature spruce in New England is in decline, but young spruce trees are thriving.

There is a great need for additional studies in forest pathology and entomology. It is now recognized that stress in tree growth is caused by a variety of factors other than acid rain, and there is a need to know how much of the variability in forest growth is associated with acid rain. Such questions must be raised as, Will the pH level of the precipitation falling on the trees have a long-term effect on the trees that may prove debilitating or disastrous? Is acid rain affecting the forests by affecting insects, disease, or soils? Or is it possible that acid rain is not responsible for any forest damage?

If it is found that air pollution is causing a decline in the growth of the trees, leading to a subsequent attack by other organisms, controls must be developed. Control is based on cost/

benefit—the costs of cleaning up the cause of the problem as opposed to the benefits of cleaning up the environment. According to industry estimates for the United States alone, the cost of cutting down on sulfur and nitrogen emissions is $200 billion. If the decision is based on economics alone, the destruction of the forests will continue. But the cost/benefit view is not the only consideration. The citizens of a nation have the responsibility to take care of the environment. There must be a balance between economics and the social welfare of the people.

Soils

Acid rain has an adverse effect on naturally acidic soils because of the behavior of ions in the soils as their acidity is increased. Most of the ions in naturally developing soils are positively charged. These include the hydrogen ions (acidity); nutrient ions, such as calcium and magnesium; and potentially toxic ions, such as aluminum, lead, mercury, and cadmium. These positively charged ions (cations) normally do not migrate through the soil because they are tightly bound to the negatively charged surface of large, immobile soil particles. The negative ions consist of silicates on the surface of clays and of organic acids on particles of organic matter. The ability of the soil to bind positively charged ions is called the cation-exchange capacity, and the chemical bonding is responsible in large part for controlling acid deposition. The continued addition of acid can destroy the cation-exchange capacity of the soil, thus increasing its acidity.

Acid rain affects the positive ions in soils in two ways. First, the hydrogen ions in acid rain displace the positive ions from their binding sites. Consequently, the hydrogen ions are increasingly concentrated in the soil water. Second, the sulfate and nitrate ions in acid rain, being negatively charged, set in motion a reaction that allows the positive ions to be leached from the soil. As a result, acid rain permits the release of hydrogen, nutrient, and toxic ions from the soil.

The negative ions, anions, are also affected by excess acidity. Such nutrients as boron, molybdenum, and phosphorus exist as negatively charged borates, molybdates, and phosphates, which plant roots take up by exchanging them for negative hydroxide and bicarbonate ions. As acidity increases in the plant tissues, there are proportionately fewer hydroxide ions available to be replaced, as well as fewer nutrient anions in the soil.

Phosphorus, which is readily available to plants that have a pH level between 5.5 and 7, reacts with metals in acidic soils to form insoluble compounds that roots cannot absorb. Without phosphorus, plants are deprived of an essential chemical aid in the transfer and storage of energy for biochemical reactions. Boron and molybdenum are only trace elements, but they are essential to plant growth. Molybdenum is needed to permit plants to utilize nitrates, and without molybdenum, the leaf structure of a plant is harmed.

A change in pH level can also affect the structure of soil grains or aggregates and their resistance to breakdown during rainfall and tillage; the rate of decomposition of organic matter by bacteria; the rate at which bacteria fix nitrogen in the soil; the growth of disease-causing fungi; and the survival of animals, such as earthworms, that aerate the soil. The germination and growth of seeds is dependent upon the structure of the soil. Stable soil grains and pores permit a rapid water intake but also good drainage, which allows the plants to make effective use of the available water. In acidic soils the aggregates cannot maintain their structure. As a consequence the soil compacts, and a hard surface crust may form, blocking the entry of both air and water.

Acidity will affect the structure of the soil in a number of ways. Calcium will satisfy the negative charge on clays and bind the particles into aggregates. With acidification the calcium ions are replaced by hydrogen and aluminum ions, which reduce the structural stability of the soil.

Another factor that affects soil structure is the organic-matter content of the soil. Microscopic soil organisms, principally bacteria, secrete complex carbon compounds that help cement aggregates together. When acidity rises this metabolism slows, and when the soil becomes highly acidic the organisms are destroyed.

In essence, soil is radically changed by acidification, and is no longer able to provide nutrients in sufficient quantities for maximum yields. In areas of high acidification the increase of calcium and phosphorus in lake waters indicates that these nutrients are being leached from the surrounding acidic areas. Concentrations of aluminum, vanadium, and other potentially toxic metals increase in acidified soils.

At the present time no definitive judgments have been made as to the ultimate effects of acid rain on soil formation over an extended period of time. There can be little doubt, however, that acid rain does have a derogatory effect on the productivity of soils.

Materials

The economic loss in the United States owing to the damage done to building and other materials by air pollutants is now estimated to be well over $300 billion, up from an estimated damage in 1949 of $1.5 billion. Equally high economic losses have been reported in other countries where acid rain has increased. Of particular importance is the deterioration of metals, carbonate building stones, paints, fabrics, rubber, leather, and paper. In many parts of the world where historical monuments have existed for ages, the air pollutant damage has been incalculable. Because these treasures are irreplaceable, their preservation from the destructive effects of acid rain poses a challenge that must not be ignored.

The corrosion of metal represents one of the most universal effects of air pollution. Of the metals in common usage, iron and steel account for about 90 percent of the total. Although corrosion is natural, pollutants accelerate the process, and sulfur dioxide is the most detrimental. When SO_2 is changed by oxidation to sulfuric acid, the rate of corrosion is increased, and moisture is critical in this process. When the relative humidity is below 60 percent, corrosion is slow; when it is above 80 percent, corrosion is very high.

Nonferrous metals also are affected by atmospheric pollutants. When SO_2 reacts with copper, the familiar green patina of copper sulfate results. Because nonferrous metals are used to form electrical connections in equipment, corrosion can cause a malfunction and result in serious operational and maintenance problems. Pollutants may also damage low-power electrical contacts in computers, communication equipment, and electronic equipment.

Of the building stones, limestone ($CaCO_3$) and dolomite ($MgCO_3$) are strongly attacked by atmospheric pollutants. In contrast, granite, gneiss, and sandstone are little affected. Sulfuric acid reacts with the calcium carbonate to produce water, carbon dioxide, and gypsum ($CaSO_4$), which appears as a white crystalline or fibrous mass on the stone surface. These erosion effects are not limited to the surface, for water moves the acid into the interior of the stones. As a result, an entire stone will eventually be washed away. A number of priceless historical monuments and works of art have been damaged or destroyed in this manner in recent times. To illustrate, in the late nineteenth century the obelisk Cleopatra's Needle was moved from Egypt to a park in New York City. Although the monument had withstood the ravages of desert heat and sand for thousands of years, the hieroglyphics have been obliterated from the needle in the last half century. In Greece, the marble of the Acropolis is dissolving; in London, St. Paul's Cathedral is being eroded; and in Washington, D.C., the marble of the Capitol is pitted with craters a quarter of an inch or more in diameter.

A host of other materials, including leather, paints, and textiles, are being attacked. In 1983 an EPA study indicated that acid deposition degraded the film base of photographic negatives and weakened the tensile strength of cotton and nylon. Paint weathering is accelerated by pollutants, which may cause soiling, discoloration, loss of gloss, decreased adhesion, and increased drying time. If a paint coating is sufficiently damaged, the underlying surface is open to attack by pollutants. A light paint may be dissolved when lead sulfide, a black substance, forms. The discoloration depends not only on the concentration and the duration of exposure but also on the lead content of the paint.

Sulfide and nitrogen compounds in the air are so pervasive in many urban areas that numerous libraries, including the Library of Congress in Washington, D.C., and Chicago's Newberry Library, have found it necessary to install small-scale atmospheric "scrubbers" to filter pollutants from the incoming air to protect the valuable and, sometimes, rare books. Microfilmed records of historic documents and archives are particularly vulnerable to the effects of pollutants.

Human Health

There is growing evidence that acid rain proves a threat to human health. As long ago as 1872, R. A. Smith, a chemist in Glasgow, Scotland, attributed the "great mortality" in the city to the high levels of acid sulfate in the atmosphere. When such an accident occurs it is easy to locate the cause, but it has been remarkably difficult to quantify the long-term health effects of low-level acid rain.

A number of studies have attempted to show relationships between acid rain and human health. The New York State Department of Health has linked acid rain to an outbreak of gastroenteritis in the Adirondack Mountains—the outbreak was caused by acid-resistant strains of bacteria found in wells fed by acidified groundwater. In another report the department found that water from acidified lakes contained toxic chemicals that were potentially dangerous to health and recommended that the residents of affected areas flush their water pipes thoroughly before drinking any more water.

Studies have shown that sulfur and nitric oxides pose serious problems for people with respiratory diseases, and the presence of sulphate particles in acid rain can lead to serious illness and even death. The U.S. Council on Environmental Quality has estimated that "the health related problems directly attributable to acid precipitation are costing the country $2 billion annually." In a Canadian study, undertaken at Brookhaven National Laboratory in New York, it was estimated that 5,000 Canadians may be dying each year of diseases directly related to acid rain. Many cases of chronic bronchitis and emphysema leading to chronic heart disease can be traced to the effects of acidic particulate matter.

Toxic Metals and Oxides

There is growing concern that acid rain is leaching mercury, lead, cadmium, and other toxic metals from soils and bedrock. The leached metals then become available to humans in their drinking water and enter into the food chain. A low concentration of these metals can damage the nervous system of individuals within a very short period of time. To date there have been no reports of acute metal poisoning attributable to acid rain, but toxic metal poisioning is cumulative and could cause a chronic health problem in the future.

Of more immediate concern is the fact that acidic drinking water can leach lead and copper from pipes, and only very limited research has been done on this problem thus far. In a study in the Adirondack Mountains health officials in New York State found at least two cases of elevated lead levels in the blood of children whose water supply had pHs of 4.2–5.0. New York State's Environmental Health Center reported: "Natural spring water in certain locations can contain copper up to 1 mg/liter and lead up to 0.2 mg/liter. In other words, concentrations of copper and lead equal to the U.S. drinking water standards can be leached by acid rainwater from natural rock and soil formations alone." It has been found that six counties in New York State have had drinking water with lead levels that exceed federal standards.

In the sampling of drinking water in many other locations the level of toxic metals also exceeded federal standards. In a comprehensive survey of New York and New England it was found that when water stood in lead pipes overnight, lead levels exceeded federal standards in 8 percent of the households. In western Pennsylvania one county that received heavy doses of acid rain had levels exceeding federal standards in 16 percent of its households. In more than 40 percent of the surface waters in New England and New York the concentration of aluminum was ten times higher than the limit of aluminum in kidney-dialysis water, and a high level of aluminum has been found in the brains of patients with Alzheimer's disease.

Many toxic metals are emitted into the atmosphere during the combustion of fossil fuels, including lead, mercury, and cadmium, and these pollutants are now accumulating in the environment. When the sediments at the bottom of acidified lakes are analyzed, the level of trace metals has increased three- to fourfold since 1960. These toxic metals have found their way into the food chain, and fish that contain mercury exceeding the federal standards have been found in acidified lakes of New York State, Maine, Canada, and Scandinavia. Even a small increase in acidity increases the potential for toxic metal poisoning because the trace metals, such as mercury, become more soluble and their concentration is increased. Mercury is particularly dangerous, for when the acidity is higher, microorganisms automatically convert it to its most toxic form.

Nitrogen dioxide (NO_2) has a much higher toxicity than sulfur dioxide (SO_2). Elevated levels of NO_2 occur in regions where there is high traffic density, the manufacturing of nitric acid, electric arc welding, and the use of explosives; it is also found in farm silos. Because it is not readily soluble it penetrates deep into the lungs to cause tissue damage. When the concentration of NO_2 is high, abnormal pathological and/or physiological changes can occur. The pathological changes include destruction of cilia, alveolar tissue disruption, and obstruction of respiratory bronchioles. High levels of NO_2 may also aggravate respiratory infections, and there is further evidence that the respiratory mechanisms are disrupted, allowing bacteria to proliferate and enter lung tissues. In order to induce such effects nitrogen dioxide exposures must be in excess of those found under normal air conditions.

Control of Acid Rain

As atmospheric pollutants have increased, there have been proposals to reduce emissions. The reduction in sulfur dioxide emissions, particularly from coal-fired boilers, is usually considered to be the primary way to reduce acid rain. There are five principal means of reducing these acid emissions: (1) a change to nonpollutant fuels, (2) control of pollutants before combustion, (3) control during combustion, (4) reducing the emissions by scrubbing after combustion, and (5) energy conservation.

Changing to Nonpollutant Fuels

Converting to low-sulfur coal and oil is the easiest way to reduce sulfur emissions from fossil fuels, but unfortunately the amount of low-sulfur coals available is limited. Most of the coal mined in the Appalachians and the Midwest has a high sulfur content. In the United States the low-sulfur coals are concentrated on the Great Plains and in the Rocky Mountains. The regions with the world's largest reserves of oil—the Middle East (55 percent) and Latin America (about 11 percent)—produce crude oils with a high sulfur content. Shifting to low-sulfur fuels raises such important questions as, What are the regional economic impacts? What employment shifts are regional in the fuel industries? and What is the ability of the fuel industries to accommodate such spatial changes?

Converting to lower-sulfur fuels is often the cheapest way to reduce sulfur dioxide emissions as the purchase of technological controls is often a great deal more expensive. The net effect of fuel conversion by many consumers could lead to shifts in production and employment away from the traditional higher-sulfur coal regions.

Control before Combustion

The amount of sulfur and nitric oxides given off by coal can be reduced by application of physical and chemical processes before combustion. Coal is normally cleaned at the mine to remove particles of sand, clay, and other impurities. Coal washing is a relatively simple process in which the coal is crushed and the particles put through large tanks of water. In this process 50–90 percent of the pyrite is removed from the coal, but removing the organic sulfur involves a much more complex and expensive chemical process, based on the use of microwave energy and electron beams. Chemical treatment can remove both pyritic and organic sulfurs, but none of the processes has reached the commercial stage of development.

A process that shows promise for the future in the removal of pyritic sulfur from crushed coal is the dry electrostatic process. When coal is fed into a rotating drum and electrostatically charged, the coal adheres to the drum, and the pyrites and ash are removed. Experiments indicate that this process, when perfected, will remove from 38 to 68 percent of the sulfur and from 50 to 60 percent of the ash.

Coal cleaning is now widely practiced throughout the world, and about 40 percent of the coal produced in the United States is cleaned. As a result it is estimated that sulfur dioxide emissions are reduced annually by 2.4 million tons. If all the coal were cleaned, sulfur emissions in the United States would be reduced by more than 5 million tons.

When high-sulfur oils are refined most of the sulfur is concentrated in the heavy oils, whose sulfur content is usually 3.5–5 percent. The sulfur content of these heavy oils can be reduced to as little as 0.5 percent by commercially feasible techniques. In this process the heavy oils are distilled again, and the product of this second distillation is "hydrotreated" so that the sulfur reacts with the hydrogen. This hydrotreated distillate is then blended with a light oil to yield a low-sulfur fuel oil product. This process can reduce the original sulfur content by 30–42 percent.

Direct desulfurization, which is potentially more effective, is also now commercially feasible. In this method the residue from the initial refining is reacted directly with hydrogen and then reblended with the distillate to produce a lower-sulfur oil. This process has the potential of removing 90 percent of the sulfur from the oil.

Although the techniques are simple and the costs minimal, the desulfurization of heavy oils has lagged in recent years. The declining demand for fuel oils and the world oil surplus (including the availability of naturally low-sulfur oils) have led to a reluctance on the part of oil companies to invest in desulfurizing equipment. In addition, few European countries, except Sweden, have stringent regulations on the sulfur content of heavy fuel oils.

Control during Combustion

A number of processes have been developed to reduce the amount of sulfur and nitrogen compounds released during the burning of fuels. In the conventional furnace fuel is burned in a fixed bed or by suspension firing. In the process known as fluidized bed combustion (FBC) the fuel is burned on a perforated bed containing mineral matter, usually limestone or dolomite, and residual ash from previously burned coal. The coal and mineral matter are held in turbulent suspension by the upward flow of air, hence the term *fluidized bed.* The combination takes place without flames at temperatures of 815°–899° C, much less than the 1,649° C required in the conventional coal-fired system. The lower temperature results in lower nitric oxide emissions. About 90 percent of the sulfur dioxide is also removed in this process. The heat converts the limestone to lime, which reacts with the SO_2, water, and oxygen to produce calcium sulfite and calcium sulfate. The process can continue indefinitely as long as the limestone/dolomite is replaced. Although this technique was developed in the early 1970s to replace the conventional furnace, its use has been limited.

Another process, known as lime injection in multistage burners (LIMB), is still in the developmental stage. When a mixture of powdered coal and limestone is injected into the firebox and the combustion temperature is lowered, using special burners, the limestone reacts with the SO_2 to form calcium sulphate dihydrate (gypsum). This process can remove up to 80 percent of the SO_2 and about 50 percent of the NO_X emissions. The equipment can be retrofitted into existing plants, and its cost is far lower than that for flue gas desulfurization, a postcombustion cleaning technique. A simplified version of the LIMB process is used in several European countries where brown coal is mixed with limestone before combustion to reduce SO_2 emissions.

Postcombustion Control

As the flue gases escape through the stack the SO_2 and NO_X pollutants and particulates can be removed by a number of techniques. The flue gas desulfurization (FGD) method has been available for many years and is widely used in all parts of the world to reduce SO_2 emissions. Over 50 FGD processes have been developed since 1980, but only a few have been employed in full-scale operations. In FGD the SO_2 is removed after combustion by "scrubbing" the flue gases in the chimney of the stack with a chemical absorbent such as lime or limestone. There are two types of scrubbers, wet and dry.

The wet process involves the wetting of the particles by the scrubbing liquid. The particle is trapped as it travels from the supporting gaseous state across the face of the liquid scrubbing medium. In a spray chamber droplets are sprayed through the gases so that the particles are held in the bubbles; the smaller the droplets, the greater the surface area for a given weight of liquid and the greater the probability of wetting the particles.

Two types of wet scrubbers are currently used. In the first type, which utilizes a regenerable system, a chemical solvent removes the SO_2 from the scrubbing medium, and the solvent is recirculated within the system. Depending upon which regeneration technology is employed, commercial products can also be recovered, such as elemental sulfur, sulfuric acid, liquefied SO_2, and gypsum.

In the nonrecovery or "throw away" system, lime, limestone, sodium alkali, or diluted sulfuric acid is utilized. This system is widely used because it is inexpensive and removes from 90 to 95 percent of the SO_2 formed during combustion. The major problem of the nonrecovery system is that it produces a large quantity of waste materials. This type of scrubbing produces calcium sulfite and calcium sulfate, which have the consistency of toothpaste, and this sludge is difficult to dehydrate and store. In a typical 1,000-MW electric utility plant that burns coal with 3.5 percent sulfur, this process produces about 225,000 tons of sludge annually. The wet FGD process also raises the problem of how to prevent toxic metals like mercury from leaking out of the sludge and contaminating the groundwater. One major advantage of the process is that sulfur and sulfuric acid can be extracted from the sludge.

For the dry scrubbing technique two different processes are available. A lime/limestone slurry is sprayed into the flue gases and absorbs the SO_2 to produce calcium sulfate, or ammonia is sprayed into the gases to produce pelletized ammonium sulfate, which can be used as fertilizer, as a dry end product. Dry scrubbing will remove from 70 to 98 percent of the SO_2 and produce a waste product that is easier to handle than that from the wet systems. The dry scrubbers are most effective with low- or medium-sulfur coals, and commercial plants are in operation in Europe.

Several experimental systems have been devised to remove NO_X in a postcombustion process. About 50 systems have been tried, and a few of these are being used in Japan in oil-fired boilers. Most commercial flue gas denitrification methods involve selective catalytic reduction (SCR), by adding ammonia to the furnace exhaust gases and passing the mixture over a fixed-bed catalyst. This process removes about 80–90 percent of the NO_X by converting it into nitrogen, but it is expensive and has had little utilization. NO_X reduction has also been achieved with noncatalytic reduction by injecting ammonia directly into the combustion ozone, which removes 40–70 percent of the NO_X.

There is one process in the pilot stage known as the electron beam method that removes SO_2 and NO_X simultaneously. The flue gas is first sprayed with water, to reduce its temperature, and then with ammonia. The cooled gas mixture than passes through an electron beam reactor, which turns the oxides into acids. These form ammonium compounds that are used as fertilizer. Laboratory tests have shown that this process removes 80–92 percent of the SO_2 and 50–90 percent of the NO_X, and the U.S. Department of Energy began a program in 1984 to determine if the technique is practical.

Energy Conservation

A major way to reduce SO_2 and NO_X emissions into the atmosphere is to develop energy conservation policies. Although there is much discussion of energy conservation during times of energy emergencies, the programs are rarely implemented. The developed countries, which use more than 80 percent of the world's commercial energy, still rely on methods that were developed during an era of cheap and abundant energy. To develop a conservation program in order to reduce energy consumption, there must be improved energy efficiency, increased use of public transportation, and reduced motor-vehicle speed limits. As the consumption of fossil fuels declined, there would be a corresponding reduction of emissions from fossil fuels. In the underdeveloped countries of the world energy conservation is even more important in order to reduce the cost of imported fuels.

Carbon Dioxide (CO_2)

Carbon dioxide is one of the important trace elements in the atmosphere. Although it has a concentration of only about .03 percent by volume, a small change in the amount may play a critical role in altering the climate of the earth. With the increased consumption of fossil fuels and the destruction of the forests in the past two centuries, great quantities of carbon dioxide have been emitted into the atmosphere. Although there is no certainty that the earth's climate will be altered because of an increase in carbon dioxide, the potential for change is great.

The importance of carbon dioxide in certain amounts is difficult to overstate. First, the carbon in carbon dioxide is what makes all plant and animal life on the earth possible, and the photosynthesis of plants would not be possible without carbon dioxide in the air. Second, if there were no photosynthesis there would be no oxygen in the atmosphere as oxygen originates almost solely from the breakdown of carbon dioxide by plants. Third, the presence of carbon dioxide in the atmosphere makes the earth livable. There is liquid water on the earth because the average temperature is high enough; if there was no carbon dioxide in the air, it would be likely that all water would be frozen because the average temperature of the earth would be well below the freezing point. Carbon dioxide is an essential part of the atmosphere. The basic question is, What happens if its concentration is altered?

Effect of Carbon Dioxide on Energy Distribution

Since the 1960s the "greenhouse effect" of carbon dioxide has been increasingly recognized. In a greenhouse, the sun's rays enter through the windows. When the rays strike an interior surface, energy is released, and the building is warmed. The wavelength of the energy is changed from short to long. The greenhouse window passes the short waves out readily but traps the long waves, thus trapping heat in the greenhouse.

Within the atmosphere certain substances act in the same manner, allowing sunlight in but preventing some of the heat from radiating back into space. Water vapor is one of these substances, which explains why humid nights cool off much more slowly than dry, clear nights. Carbon dioxide acts in the same manner.

Therefore, there is an initial assumption that an increase in carbon dioxide should warm the earth. Some spectacular stories have been written to show that the temperature of the earth will increase so that the deserts will be enlarged, the polar ice caps will melt, and the coastal areas of the world will be flooded. Before such an assumption can be considered conclusive, there needs to be a careful study of just how carbon dioxide can change the atmosphere. Carbon dioxide cannot change the amount of energy coming from the sun, and it does not reflect energy back into space before it enters the atmosphere. Therefore, it cannot change the amount of energy in the total atmospheric system; it changes the distribution of the energy within the system. In reality it can raise the temperature of the atmosphere near the ground and lower the temperature at higher altitudes.

It has sometimes been stated that carbon dioxide can change the reflectivity of the sun's incoming rays by leading to an increase in the cloud covering. If this increase were to occur there could be a change in the amount of energy reaching the earth from the sun. Photographs from space demonstrate that clouds do reflect energy back into space. If the carbon dioxide content of the atmosphere were to change the cloud covering significantly, the earth's overall temperature could be lowered. But it is now known that there is some feedback, that is, some self-regulation, in the process of cloud formation. Greater cloudiness cools the earth somewhat initially, but in the process less water evaporates from the water and land and cloudiness decreases. Recent studies have shown that this self-regulating mechanism prevents changes in average cloudiness and thus prevents significant changes in average world temperatures.

Atmospheric Content of Carbon Dioxide

The carbon dioxide content of the atmosphere was essentially stable until about 1860 when the great increase in fossil fuel consumption began to alter carbon dioxide content of the atmosphere. The present measurements of the changes were begun by C. D. Keeling of the Scripps Institution of Oceanography in 1958. A carbon dioxide monitoring station was placed on Mauna Loa in Hawaii, and this station provides data on the carbon dioxide content of the troposphere, or lower atmosphere.

The data from Mauna Loa and other more recently established stations indicate that there has been a steady increase in the amount of carbon dioxide in the atmosphere. The amount of the increase at Mauna Loa is about 0.8 parts per million. Extrapolating back to 1860, the amount of carbon dioxide in the air has increased from 290 parts per million to more than 330 parts per million. About 25 percent of this total increase has come since 1965. If this rate of increase were to continue to the year 2220, there would then be approximately twice the amount of carbon dioxide in the atmosphere as there is now, or 235–275 thousand million tons of carbon dioxide. The present rate of increase is approximately 8–9 thousand million tons, or 1 part per million, each year.

Not only has there been a steady increase in the amount of carbon dioxide in the atmosphere, but the amount varies seasonally. In the Northern Hemisphere, the carbon dioxide content rises to a peak in late winter or early spring and falls to a minimum in late summer and early fall. The amount of carbon dioxide in the atmosphere will thus vary seasonally from about 0.5 parts to 1.5 parts per million. These seasonal changes are directly related to the process of photosynthesis that occurs during the summer months. The forests are most important in this increase in carbon dioxide because of their vastness worldwide and their potential for storing carbon in quantities that are sufficiently large to affect the carbon dioxide content of the atmosphere. The differences range from about 15 parts per million in the northeastern part of the United States, where forests exist, to 5 parts per million at Mauna Loa in the Pacific Ocean. The differences from season to season drop toward the tropics, where seasonal changes are slight. The release of carbon dioxide from the biota has been altered through the destruction of forests and the oxidation of humus, and the assumption that the increase in carbon dioxide content of the atmosphere is the result of the burning of fossil fuels without regard to change in the forest covering of the world is incorrect.

Recent studies show that the world carbon budget is more complex than was originally thought. It is now known that there are a number of "pools" of carbon and that there is a more or less continuous interchange of carbon among them. The atmosphere at present holds about 700×10^{15} grams of carbon in the form of CO_2 (approximately 700 billion tons of carbon in the atmosphere), which is continuously being exchanged with the biota and the surface waters of the earth. The amount of carbon worldwide in the biota is about 800×10^{15} grams. In addition there are between $1,000 \times 10^{15}$ and $3,000 \times 10^{15}$ grams of carbon in the organic matter of the soil, mainly humus and peat. The change in the area of forests, or an increase or a decrease of agriculture, would change the carbon content of the atmosphere. Finally, the largest reservoir of carbon is found in the oceans in the form of dissolved CO_2, which is a part of the carbonate-bicarbonate system. The total in this pool is about $40,000 \times 10^{15}$ grams.

The basic question is, Can any one or any combination of these pools absorb the increase of carbon dioxide in the atmosphere? At the present time the most accurately known figure for the release of CO_2 from the combustion of fossil fuels is about 5×10^{15} grams of carbon per year, and the increase in the CO_2 content of the air is equivalent to about 2.3×10^{15} grams of carbon per year, which leaves about 2.7×10^{15} grams to be removed from the atmosphere by some combination of biota and oceanic processes.

In addition to the CO_2 added to the atmosphere by the burning of fossil fuels, another 4×10^{15} to 8×10^{15} grams may be being released through the destruction of forests and accelerated oxidation of humus. The combined figure is thus an increase of 9×10^{15} to 13×10^{15} grams of carbon per year. With only 2.3×10^{15} grams stored in the atmosphere, the major question is, Where are the remaining 6.7×10^{15} to 10.7×10^{15} grams of carbon stored?

This is a major unanswered question. Scientific studies indicate that the oceans cannot act as pools for large additional amounts of carbon, and scientists are reviewing their original assumptions to determine if some types of mechanisms are present, which have been previously ignored, that can remove additional carbon from the atmosphere. Nevertheless, there is evidence that much more scientific work is needed in order to understand the types of mechanisms that can change the world's carbon budget. These studies include the acquiring of improved knowledge about the changes in the spatial distribution and structure of forests worldwide and the investigation of biotic mechanisms that will increase the knowledge of how those mechanisms might increase the transfer of considerably more carbon into the ocean depths than is at present thought to occur. The immediate prospects for developing these types of studies do not appear to be promising because of budgetary limitations in government and university research centers.

Potential Effects of CO_2 on the Climate

The probable changes in average annual surface temperatures and precipitation levels at different latitudes have been estimated for an increase of atmospheric carbon dioxide to 620 (plus or minus 60) parts per million, somewhat less than twice the present level. The results are shown in Table 1.

These increases in temperature would have a profound effect on the precipitation over the entire world. If these changes do occur, there will be different spatial allocation of water, and as shown in Table 1, certain areas will have a possible excess and others a deficit. Such changes have immense potential for political and economic repercussions.

TABLE 1

Probable Changes in Average Annual Surface Temperatures and Precipitation Levels if Amount of CO_2 in the Atmosphere Increases to 620 Parts per Million

LATITUDE	AVERAGE ANNUAL CHANGE IN SURFACE TEMPERATURE (IN ° C)	AVERAGE ANNUAL CHANGE IN PRECIPITATION
60° N	+7.5	+18
50° N	+6.0	+ 4
40° N	+6.0	−14
30° N	+4.5	0
20° N	+2.5	+20
10° N	+1.5	+20
Equator	+3.0	0
10° S	+4.0	−20
20° S	+4.5	− 5
30° S	+4.0	+ 5
40° S	+4.0	+12
50° S	+3.0	+12
60° S	+2.5	+12

Source: *Carbon Dioxide and Climate: The Greenhouse Effect,* Hearings, Committee on Science and Technology, U.S. House of Representatives, 97th Cong., 1st sess., July 31, 1981, p. 22.

There is currently no measurable evidence that the climate has changed because of an increase in the carbon dioxide content of the atmosphere. It is not likely that the temperature variation from year to year of three-tenths to four-tenths of 1° C is the result of the carbon dioxide effect, but although the problem of climatic change does not appear to be present today, it is not rational to ignore the problem until a major climatic change does evolve.

At present there is a considerable body of information about the amount of CO_2 in the atmosphere, much of it qualitative in nature, but the information is still inadequate for finite discussion. Present-day information does reveal that (1) the problem is global, (2) carbon dioxide change in the atmosphere could reach a level that has not yet been experienced, (3) the problem is long range, and (4) it is closely related to the global energy system.

In order to develop a program to reduce CO_2 emissions into the atmosphere, all countries of the world must be involved, industrial as well as Third World countries, for the emission of carbon dioxide through the burning of fossil fuels and wood is universal. There must be universal agreement for a control of those emissions, and there must be a cost/benefit analysis to show how the control of carbon dioxide will affect the economies of the world.

A large number of scientists believe that the critical bulk of carbon dioxide sufficient to change the world's climate will not occur until well into the twenty-first century. However, the decisions that are made in the immediate future will determine the amount of carbon dioxide emissions and will involve such critical problems as, How much fossil fuel will be burned? How much nuclear power should be developed? and How much of the world's forests should be cut? If we err too much in increasing the use of fossil fuels and the burning of wood, the environmental consequences could be substantial. If, on the other hand, we require the expenditure of huge sums of money for overly stringent pollution protection equipment, the situation may be just as catastrophic from an economic viewpoint.

In order to provide information so that a logical program can evolve, a number of basic scientific studies are critical. A fundamental question is, How much carbon dioxide will stay in the atmosphere and how much is deposited in the carbon pools? The Committee of Science and Technology of the U.S. House of Representatives recognized in 1968 that three kinds of research were needed to (1) assess the risks that can be averted only by limiting carbon dioxide emissions, (2) evaluate beneficial effects of carbon dioxide in the atmosphere and how to lessen harmful ones, and (3) study potential social and institutional responses to the consequences of more atmospheric carbon dioxide and projected climatic changes.

Assessment of Risks

The studies of risk assessment would include both physical and human problems. One investigation that needs to be carried out concerns the possible disappearance of the West Antarctic ice sheet. Many glaciologists believe that the Antarctic ice cap is unstable, and if there were a worldwide warming and this ice cap were to melt, many coastal cities would be inundated and much valuable agricultural land would be lost. There is a need to estimate the probability of this event and to approximate whether it would occur in decades or over centuries. As an analogue to this investigation another should investigate whether the West Antarctic ice cap disappeared during the last interglacial period.

Studies also need to be made on the effect of warming on the Arctic Ocean's ice. A recent model indicates that a fourfold increase in atmospheric CO_2 would raise temperatures sufficiently to melt the Arctic ice, which would cause profound changes in oceanic and atmospheric circulation. To verify the model we need a greater understanding of the dynamics of the formation and dissolution of sea ice.

The potential climatic changes would affect not only the oceans, but also the land masses. In the Northern Hemisphere a warming of 5°-7.5° C might occur over land in the higher latitudes, which could result in the thawing of the permafrost and thus affect the hydrology and the ecology of the tundra. The tundra contains large quantities of organic carbon and methane in peat deposits, and the thawing of the permafrost could result in an oxidation of the peat and a release of carbon dioxide and methane gas—which would further add to the amount of CO_2 in the atmosphere.

There is major concern about the possibility of increased atmospheric warming that would be difficult or impossible to modify by human actions. This problem involves the probability of determining the time of such an event and such questions as, What would be the effect of CO_2 concentrations on human respiration and could this affect life expectancy? If summer temperatures were to increase, could life-threatening stresses affect certain groups of human populations? and If the upper atmospheric temperature were to decrease, how would this fact interact with additions of chlorofluorocarbons and other gases to change the ozone content of the stratosphere?

Evaluation of Benefits

Studies to enhance the beneficial effects of increased carbon dioxide in the atmosphere and lessen harmful ones have the objective of increasing productivity in agriculture, forestry, and animal husbandry. Such questions must be raised as to what would be the biological effects of increased carbon dioxide on crop plants? There is a possibility that the increase in crop yields in the twentieth century may in a small way be related to the increase in the carbon dioxide content of the atmosphere. There should also be research on the effects of higher temperatures and increased CO_2 on biological nitrogen fixation and on weeds, insects, and microbiological pests.

An increase in atmospheric carbon dioxide could produce environmental stresses in many crops because of such factors as higher temperatures, lower water availability, increased pollutants, and increased soil alkalinity or higher soil salinity. To investigate these potential problems, we need studies on the basic biology of plants and their response to environmental stresses, the genetic manipulation of crop varieties to produce crops with greater resistance to environmental stresses, and the development of production-management systems to maximize the evaluation of stress resistance plants. The ultimate objective is to increase average crop yields over an extended period of time.

Needed research has begun to show that increased atmospheric carbon dioxide can raise forest yields. In many areas of the world, but particularly in the low-latitude regions, energy from the burning of wood can be substituted for coal and petroleum. However, this substitution of biomass energy can occur only if forest yields are maintained and if possible increased.

The effect of a climatic change on animal productivity could be disastrous. It is a well-established fact that with present animal varieties, nutrient intake and reproductive rates diminish with rising temperatures. Physiological research on the metabolic and endocrine systems of domestic animals under conditions of increasing average temperatures and long, hot summers is needed to understand and quantify these effects. These studies should lead to improved management practices, including better animal feeding, synchronizing breeding periods with seasonal climatic changes, and the development of low-cost animal shelters to buffer climatic change.

The model showing the potential changes that might result from increased atmospheric carbon dioxide indicates that vast areas will become drier, and the traditional agricultural systems will be especially vulnerable to such a change. Studies are needed on water regions to prepare for these possible changes. In an agricultural society the dependability of irrigation water is essential, and in many river basins an increase in irrigation systems would require the utilization of both surface- and groundwater resources. Changes in the cropping patterns, times of planting, and area planted would also be of importance in maintaining the economy of a region.

Social and Institutional Responses

If there is a major change in the climate, there will also be societal and institutional responses, and many such problems can be anticipated by the development of simulation models. The objective of these models should be to create plausible and consistent combinations of assumptions and estimates about a regional climate and its effect on the economy and society's responses. One model could be devised that would attempt to predict future international behavior; another model might reflect the self-interests of a particular nation. The process of model building should be iterative, with each successive round aimed at achieving greater consistency and sharper definitions of unanswered questions.

The changes in global or regional climates will affect each society differently, and one way to approach this complex situation is to develop case studies showing the possible impact of a climatic change. Historical case studies of climatically vulnerable areas such as the Great Plains of the United States or the Sahel in central Africa may be useful in predicting how societies adjust to changing physical conditions.

Stratospheric Ozone

The availability of ozone (O_3) in the atmosphere is critical to the maintenance of life on earth. Although small quantities of ozone may occur close to the earth in the troposphere, it is concentrated in the stratosphere from 30 to 60 miles above the earth.

Ozone is formed in the stratosphere when ultraviolet radiation splits diatomic molecules of oxygen (O_2) into two atoms. The two atoms (O) then combine with two diatomic molecules of oxygen (O_2) to produce two molecules of ozone (O_3). These ozone molecules are then broken down by ultraviolet radiation to form the original diatomic molecules of oxygen (O_2) and the atoms of oxygen (O). Under natural conditions there is an equilibrium between the creation and the destruction of ozone (O_3).

Ozone absorbs a large percentage of the ultraviolet radiation emitted by the sun that is harmful to humans, animals, and plants. Some ultraviolet wavelengths—normally referred to as UV-B—are critical to life. The concentration of ozone at different altitudes can affect the movement of ultraviolet rays through the atmosphere, which in turn influences the radiative and meteorological processes that determine weather conditions. Thus, if the ozone equilibrium is disturbed, major environmental changes could occur.

Ozone Depletion

In the mid-1970s Mario Molina and F. Sherwood Rowland produced models that indicated that the emission of oxides of chlorine, nitrogen, bromine, and other compounds into the atmosphere could destroy the natural balance of ozone in the stratosphere. Of the chemicals emitted the chlorofluorocarbons (CFCs), which are widely used in aerosol sprays, foam blowing, refrigeration gas, and other ways, could be most critical. The CFCs have a life span of centuries, and because of their stability, they rise into the stratosphere. Once in the stratosphere, at elevations of 30–60 miles, ultraviolet radiation causes their destruction. In this process, known as photolysis, chlorine is released. The chlorine then reacts with the ozone to cause its destruction. As a result the natural balance of ozone is destroyed.

Besides the chlorofluorocarbons there are a number of other chemicals that threaten the ozone layer. Of these, halons, the chemical gases used in fire extinguishers, may be even more destructive to ozone than chlorine. The halons contain bromine, and their life span is as long as that of the CFCs. Another potential source of ozone depletion is nitrous oxide (N_2O), a chemical produced in the combustion chambers of the supersonic aircraft that fly in the stratosphere. Although the life span of N_2O is relatively short, its direct emission into the stratosphere makes it an immediate threat to ozone. Other chemicals that are found in minute quantities in the atmosphere that are a potential danger to ozone include methyl chloroform (CH_3CCl_3) and carbon tetrachloride (CCl_4).

The importance of the depletion of the ozone has been debated since the mid-1970s. These discussions have waxed and waned over the years, but a number of events in recent years have increased concern about the importance of maintaining nature's balance of ozone, and recent scientific discoveries have increased the urgency of having government controls. In 1985, British scientists reported finding losses of ozone in the Antarctic that were far greater than the existing atmospheric models could explain. This "hole" in the ozone layer was about the size of the continental United States. Although the reason for the hole is unknown and its importance is not understood, the Antarctic hole has dramatically altered government considerations because of the potential for large, unanticipated atmospheric changes. It now appears that there could be a sudden threshold effect in the atmosphere rather than a gradual incremental change.

As the scientific evidence has grown, the relationship between the greenhouse effect caused by CO_2 and the growing importance of the so-called non-CO_2 greenhouse gases, including the chlorofluorocarbons and tropospheric ozone, has become evident. With a potential warming equivalent to doubling the CO_2 content of the atmosphere, the greenhouse question has shifted from that of pure research to that of policy analysis. The scientists are beginning to recognize that there are many chemical overlaps and feedback loops in the atmospheric processes, and as a consequence, it is increasingly evident that the depletion of the stratospheric ozone is only one aspect of the broader problem of total climatic change.

Effects of Stratospheric Ozone Depletion

The consequences of ozone modification are not completely understood, but what is known provides sufficient information to give concern. The most clearly established human health effect of ozone depletion is an increase in the incidence of skin cancer in white-skinned populations. Most of these cancers are nonmalignant, but sunlight also has been known to produce malignant melanoma, a rare but frequently fatal cancer. It has been estimated that if there is a 1 percent increase in the UV-B flux—that is, in the ultraviolet wavelength between 240 and 329 nanometers (nm)—malignant melanoma mortality would probably increase by 0.8-1.5 percent. The Environmental Protection Agency estimates that constant CFC growth of 2.5 percent per year would cause an additional million cases of skin cancer and 20,000 deaths over the lifetime of the existing U.S. population.

A number of other studies reveal possible health effects. An EPA report concluded that ozone depletion may increase the incidence of Herpes virus infections and parasite infections on the skin of people, and there is an indication that increasing the UV-B flux would expand smog problems in some urban areas. The latter relates the intensity of the UV-B flux to the photolysis of formaldehyde, a product of incomplete combustion, which triggers the formation of "radicals" that generate photochemical smog. One model predicted that smog would increase by 30 percent in Philadelphia and Nashville if stratospheric ozone were decreased by 33 percent and the temperature increased by 4° C. It also predicted that ozone would develop earlier in the day, which would provide a greater opportunity for a larger spatial distribution of smog.

An economically important effect of atmospheric ozone depletion is the accelerated degradation of some plastics and paints. It is estimated that within the next century the excess damage to polyvinyl chloride could be $5 billion, a figure that could be reduced by the development of improved chemical stabilizers. Little is known about the effects of ozone depletion on plants, but recent studies indicate that about two-thirds of over 200 plants tested showed some sensitivity when UV-B was increased, and field research on soybeans has shown that yields could be reduced by about 25 percent if UV-B is increased.

The depletion of ozone in the stratosphere could trigger climatic changes, including changes in atmospheric temperatures and water vapor concentrations. In essence, changes in ozone could be intimately linked with the greenhouse effect. In July 1986 the World Meteorological Organization, the International Council of Scientific Unions, and the United Nations Environment Programme concluded that "both with regard to future scientific research efforts as well as the analysis of possible societal response . . . these two environmental problems should be addressed as one combined problem."

Chlorofluorocarbon Control

The use of CFCs has grown rapidly since 1960 with respect to both amount and world distribution. Although the amount used in each country varies greatly, the average annual increase in the world was 13 percent from 1958 to 1984. Such a growth rate could extend into the future, for raw materials are completely adequate for growth of production.

The emission of the CFCs into the atmosphere varies greatly depending upon their use. Aerosols release emissions immediately; in contrast, the emissions from solid foam are released only when the foam deteriorates. The emissions can be reduced or eliminated through four basic changes: the stopping of production of all CFCs, which would provide absolute control; the reduction of operating losses; recovery and recycling during production or at the point of use; and the substitution of other chemicals that are less harmful to the environment.

Of those controls, one of the most effective would be to design and construct machinery that would reduce losses. For example, almost one-third of the CFCs produced is used for automobile air conditioning, of which about 30 percent is lost in routine leakage. The leakage could be greatly controlled by redesigning the valves and seals. Leakage losses from stationary refrigerators and home air conditioners is far less than from moving objects. In stationary machines the leakage largely depends upon the type of compression used.

The opportunities for the recovery and recycling of CFCs are substantial. Both of these methods are used today if the quantity is sufficient to justify costs, but economics and practicality provide a great barrier to recovering or recycling CFCs that are used in small, decentralized ways, such as in motor vehicles. For the CFCs that are used for degreasing and cleaning in large establishments, recovery is sometimes possible by using in-house distillation equipment. At the present time almost all excess CFCs used in the manufacture of foam are lost by venting slurry production, but by using carbon filtration in this process, the losses could be cut by 50 percent.

Within recent years it has been discovered that some forms of CFCs present little or no threat to the ozone layer—for example, when hydrogen is substituted for chlorine, the danger is greatly diminished. However, new designs and equipment are required in order to use many of the "soft" CFCs. In September 1986 DuPont, the largest producer of CFCs, announced that it would produce substitute CFCs in commercial quantities in five years.

There is also a strong possibility that another chemical could be substituted for the CFCs. In the United States and several other countries hydrocarbon is used in more than 90 percent of the propellants, and these aerosol substitutions have been considered highly successful. Also, some flexible foams are produced with methylene chloride, though health risks may limit the use of this toxic chemical. Substitutes are thus continually being developed.

Development of Regulatory Policies

Although there has been a great increase in the availability of information on the potential danger of CFCs to ozone, and in the development of substitutes, the reduction of the use of CFCs has been amazingly slow. In the period after 1974, when it was first hypothesized that CFCs could provide a potential danger to the ozone layer, a number of countries took unilateral action to reduce the use of CFCs. In 1976 the European Economic Community agreed to reduce aerosol use by 30 percent and to prohibit increased CFC production capacity. Although these endeavors appeared to reduce CFC emissions and risks to the ozone layer for several years, continued growth in many nations worked against solution of the problem.

In 1980 a number of countries were considering stronger measures of control. The Environmental Protection Agency proposed limiting CFC production to its current level and, further, allocating the allowable production through purchased permits that would gradually force a reduction in output. These measures were never implemented, for both political and scientific reasons. The administration that took office in 1981 looked unfavorably on imposing more regulations, and many scientists questioned the seriousness of the situation. It was felt that nature would maintain the natural balance of ozone in the stratosphere.

Although the United States delayed action, international discussions continued. In 1980 the United Nations Environment Programme Governing Council recommended that national governments reduce the production and use of CFCs. In 1981 the same body established an ad hoc working group of legal and technical experts to establish a global framework for the protection of the ozone layer. After several years of work, the Vienna Convention for the Protection of the Ozone Layer was signed in March 1985 by 20 countries.

The convention, consisting of some 21 articles and two technical annexes, provides the structure to control activities that "have or are likely to have an adverse effect" on the ozone layer and encourages research as well as the exchange of information. The convention went into effect in 1986, by which time at least 28 countries had signed the agreement, including the major producers and users of CFCs.

During the 1980s a number of proposals have been presented to control the production and use of CFCs in individual countries. In 1983 Norway, Finland, and Sweden proposed a protocol for controlling CFCs. Later that year the United States, Canada, and Switzerland proposed limiting the protocol to an international aerosol ban, which also became the Nordic countries' position— these countries had already banned the use of CFCs in aerosols. In contrast, the European Economic Community proposed an alternative protocol modeled after its own policy of a 30 percent reduction in aerosol use and a cap on future CFC production capacity.

Although various compromise positions have been proposed, a quick solution has not been found. In 1986 two workshops were held to review the economic and policy situation and to make certain that all nations are informed of the potential danger of ozone depletion because of CFCs. At the end of 1986, although no official action had occurred, the United States and the European Trade Association, representing CFC producers and users, agreed to support the concept of limiting CFC production. DuPont also agreed to support production limits and to develop alternative chemicals. In essence, the control of CFCs by late 1986 hinged on three questions: What are the policy implications if growth in other trace gases affects ozone depletion owing to CFCs? What is the risk of delaying controls? and What is the most effective and workable form of regulation?

After continued negotiation the representatives of 24 nations signed a treaty on September 16, 1987, that was designed to preserve the ozone layer by a reduction in the production of CFCs. The agreement would freeze world production of the CFCs in 1990 at their 1986 levels, and it would mandate a 50 percent reduction of world production of the chemicals by 1999. It would also freeze the production of halons, a related group of compounds, by 1992. The treaty was to go into effect immediately after more than 11 nations had ratified it—the draft was immediately signed by the United States, Japan, and 12 members of the European Community. The Soviet Union did not sign immediately but indicated it would do so soon. The European Community accounts for 42 percent of the production of CFCs; the United States, 33 percent; Japan, 11 percent; and the Soviet Union, 10 percent.

On December 1, 1987, the Environmental Protection Agency announced plans to freeze U.S. consumption of the chemicals at 1986 levels, then to cut back 20 percent after four years and another 30 percent six years after that. Two viewpoints continue to be discussed: Environmentalists protest that the EPA plan is insufficient, and producers of the chemicals warn that the regulations could cause a severe economic impact. Many manufacturers have predicted that the prices of CFCs will increase and that small producers will be forced out of business. Margaret Rogers of the Society of the Plastics Industry has said, "We are expecting a very severe economic impact." In contrast David Doniger of the Natural Resources Defense Council said the plan is "not far enough or fast enough." He added that the United States should be the world leader in the control of CFCs. EPA administrator Lee M. Thomas, who announced the agency's plans, has indicated that the problem is worldwide and that U.S. action, even if more drastic cuts were required immediately, would not have a significant impact. A major forward step has been the recognition that CFC emissions are a potential danger to the ozone layer, and it is hopeful that worldwide action is now occurring.

Radioactive Contamination

The potential for radioactive contamination of the atmosphere is a modern problem. On July 16, 1945, on a desert in New Mexico, the United States exploded the world's first atomic bomb, thus creating the nuclear age. With the advent of the world's first electricity generating nuclear plant in 1957, located at Shippingport, Pennsylvania, the role of nuclear power was hailed as a cheap, clean, efficient, and safe energy source for the world. The use of the nuclear bombs in military endeavors and the use of nuclear energy have been seriously questioned in more recent years. As the human consequences of radioactive contamination became apparent, so did the need to understand the devastating potential of nuclear explosions and energy.

Short-term Health Effects

A major characteristic of nuclear material is its radioactivity. The waste products are extremely complex, and the radioactive isotopes may have a half-life of less than a second to many thousands of years. Further, each radioactive isotope emits a characteristic radiation when it decays. It can be electromagnetic, such as that of X-rays, or it can produce particles such as alpha, beta, or neutron radiation. However, all radiation from radioisotopes has a similar ionizing effect when it touches matter, including biological tissue. In the ionizing process the tissue loses one or more electrons. The ionized atoms absorb a great deal of heat, and in the process chemical changes occur in the biological tissue. These chemical and ultimately structural changes can have serious health consequences.

After extensive research it has been discovered that the health effect of a massive exposure to radiation in a short period of time is quite different than the effect of exposure to about the same amount of radiation over a long period of time. For example, if individuals are exposed to 1,000 rems over a short period, immediate severe illness will occur, and almost all persons will die within one month. In contrast, if individuals are exposed to ten times the amount of radiation in the natural background over several decades, none will suffer acute illness, and only a small percentage will experience genetic defects that cause illness or cancer.

Long-term Health Effects

An exposure to ionizing radiation may not become evident for many years, but research on the delayed genetic and somatic effects is not as detailed or quantifiable as it should be. Large human populations have not been exposed to doses of ionizing radiation over extended periods, and while laboratory animals have been intensely studied, the extrapolations from animals to humans are exceedingly difficult. The long-term effects of radiation consist of three categories: genetic effects, human growth and development, and somatic effects.

The genetic effects of ionizing radiation are most easily recognized in a single, dominant mutation associated with a disease or an abnormality. About 1 percent of the population has some form of dominant mutation including anemia, dwarfism, and extra fingers or toes. It must not be thought, however, that all mutations are caused by doses of ionizing radiation. Most mutations are a response to complex hereditary causes.

In order to measure the risk of mutations caused by ionizing radiation, a method is utilized to estimate the dose that will double the naturally occurring rate of spontaneous mutations. This is referred to as the "doubling dose." Scientific studies indicate that a doubling dose of chronic radiation is between 20 and 200 rems, and since the natural background radiation is 3–5 rems in a reproductive lifetime, only a very small percentage of the mutations can be attributed to this effect (a rem is a unit of radiation dosage relating to the energy deposited in tissue and the consequent health risk from that deposition of energy). Even additional amounts of radiation will have only a small effect on increased mutations. The possible relationships between additional ionizing radiation and increased disease because of this mutational component are predicted to be between 0.5 and 5 percent in general illness for an additional dose of 5 rems per generation, and this figure may be too high. Estimates indicate that the increase in the dose to which a population will be exposed in the normal operation of all nuclear power plants by the year 2000 will be only 0.001 rems per year or 0.03 rems for a reproductive lifetime.

(Federal limits permit 3 rems per quarter not to exceed 12 rems per year with an average of 5 rems per year.)

Evidence has now been accumulated to show that ionizing radiation has a major effect on a fetus in utero and on young children. From studies conducted on individuals affected by the Nagasaki, Hiroshima, and Marshall Islands nuclear explosions, it was found that there were reduced growth rates, mental retardation, and microcephaly. Different types of doses do produce different health changes. When a pregnant mother is exposed within 16 weeks after conception, the individual is more sensitive to health problems than if exposed later. If a fetus is exposed to 25 rems, the child is mentally retarded; if the dose is 50 rems, the mental retardation is profound. If the radiation occurs after birth, much larger doses are needed over a longer period of time to cause health problems.

Unlike massive doses of radiation that produce immediate effects, the somatic effects of small doses over time may not become evident for years or even decades after exposure. The long-term effects may result in a possible decrease in fertility, but the most important effect will be the development of cancers such as leukemia or malignancies of the breast, thyroid, stomach, and intestinal tract. As a result of long-time research, it has been established that the increased risk of cancer is in proportion to the excess in radiation exposure.

Massive Radioactive Atmospheric Emissions

There are two major sources of massive emissions of radioactive materials into the atmosphere. The first is the exploding of an atomic bomb in the atmosphere. The danger of this practice was recognized in the 1960s, and international agreements have stopped this practice although atomic bombs continue to be tested underground. The second massive source of radioactive emissions is from an accident in a nuclear power plant when a reactor explodes.

Atomic Bomb Atmospheric and Environmental Contamination

The testing of atomic bombs to determine their destructiveness began in 1946, and the United States conducted 66 such tests on Bikini and Eniwetok islands in the Pacific Ocean between 1946 and 1958. The site chosen had to fulfill numerous conditions. It had to be an area controlled by the United States in a region that was free from storms and cold temperatures. It also had to be a sheltered area for anchoring target vessels and measuring the effects of radiation. Further, it had to be an area with a small population that could be readily moved to another area. The search ended with the selection of Bikini and Eniwetok; the following discussion is limited to the Bikini experience.

The Bikinians were asked if they would be willing to sacrifice their island for the welfare of humankind. After deliberating, the Bikinians agreed to move. The Bikinians chose the uninhabited island of Rongerik as their new home, but it proved infertile and food shortages soon developed. By 1948 near starvation could no longer be ignored, and new sites were explored. This time the Bikinians selected Kili, a fertile island 400 miles south of Bikini that had previously been used for copra plantations, and this island has become the permanent home for about 900 Bikinians. This drastic change from an atoll existence with abundant fish to an isolated island with no lagoon has taken a severe psychological and physical toll on the people. Although Kili is more fertile than Bikini, the Bikinians are not skilled enough in intensive agriculture to make the island productive. In 1946 they were completely self-sufficient; today they have lost the will to provide for themselves.

During the early testing the U.S. Navy remained optimistic that as soon as the testing stopped the inhabitants could return. The radioactive material, however, contaminated not only the atmosphere but also the land. The dangers of radiation had become evident by 1954, and the effects of radioactive contamination were not limited to Bikini as winds carried the radioactivity eastward to the islands of Rongelap and Utirik. In 1958 President Eisenhower declared a moratorium on U.S. atmospheric atomic testing in the Marshall Islands.

In 1969 the Atomic Energy Commission declared that Bikini had virtually no radiation left and that there was no discernible effect on plant or animal life. Many Bikinians were joyful that they could return to their island and eagerly awaited resettlement. The Department of Interior first built 40 new homes and planted 50,000 new trees on the island, and in 1973 the U.S. government indicated that construction was nearly complete. Even before resettlement, however, problems were becoming evident. In 1972 the Atomic Energy Commission found that radiation levels on Eniwetok were extremely high in certain places, and in late 1974 the secretary of the interior, Roger C.B. Morton, alarmed at the findings of routine radiological surveys, halted construction on Bikini Island. In March 1975 Defense Secretary James R. Schlesinger requested that a thorough survey be conducted of the radioactivity on Bikini, but the surveys were delayed.

Meanwhile the Bikinians had expressed a desire to live in houses in the island's interior, but a routine survey in June 1975 indicated that the interior was too radioactive for housing and that some wells were contaminated with radioactive plutonium. Further, although coconuts were likely to be safe, breadfruit and pandanus, two basic staples, contained unacceptably high levels of radiation. The Bikinians, frustrated and confused, brought suit in federal court in October 1975 to force the United States to stop the resettlement program until a comprehensive radiological survey of the island could be made. The U.S. government agreed to make the survey, and $2.6 million were appropriated in 1977, but the Defense Department took no action, stipulating that more money was needed. The squabble lasted three years. In the meantime tests showed that levels of strontium-90 in well water on Bikini exceeded acceptable U.S. standards, and medical examinations revealed that the people on Bikini had absorbed cancer-causing radioactive elements such as strontium, plutonium, and cesium.

In 1978 a medical team informed the 139 people living on Bikini that they could no longer eat locally grown food but must import all food and water. After it was discovered that the population had acquired in a single year a 75 percent increase in body burdens of radioactive cesium-137—the people of Bikini had ingested the largest amount of radiation of any known population up to that time—it was concluded that all people had to be removed from the island. In August 1978 the 139 people were removed from the island, and no one has been allowed to live there since.

The Bikini experience has raised many questions and provided some answers. Certainly there is positive evidence that radioactive contaminants deposited from the atmosphere after atomic testing have persisted for many years. There is also evidence that the U.S. government was less than responsible in its actions to the Bikinians. There remains the problem of where the Bikinians will find a permanent home. In the wider perspective, the atomic testing on Bikini in the Marshall Islands illustrates on a small scale what a modern atomic holocaust could bring.

Nuclear Plant Accidents

There have been a number of accidents at nuclear power plants since the 1950s. The first major incident was an apparent explosion of a nuclear waste dump near Chelyabinsk and Smolensk in the Soviet Union in 1958 or 1959. It is known that several villages were destroyed and that there was sufficient contamination of lakes and vegetation to make the entire area uninhabitable. However, the effects were limited to the Soviet Union, and the government did not provide information on the accident.

Although there have been a number of minor accidents in nuclear power plants, only two have received worldwide attention. On March 28, 1979, there was a partial meltdown of one reactor at Three Mile Island near Harrisburg, Pennsylvania. Although the reactor was highly damaged and contaminated the plant, there was essentially no emission of radioactive materials into the atmosphere. This accident contrasts sharply with the Chernobyl accident in the Soviet Union in 1986, when there was a complete meltdown and a massive emission of radioactive wastes into the atmosphere. Because of the importance of this accident, it is discussed at length here.

THE CHERNOBYL ACCIDENT   On Saturday, April 26, 1986, at 1:23 A.M., the number four reactor of the nuclear power station at Chernobyl in the Ukraine exploded and blew the roof off the plant. A few seconds later a second explosion ejected graphite blocks and fuel material. The hot graphite created a massive fire that threatened to engulf the power plant's three other reactors, but with herculean efforts the fire was contained to the reactor that had exploded.

A radioactive plume of materials and gases—including iodine-131, cesium-137, and strontium-90—was spewed into the stratosphere. The emissions reached a peak on April 26 when about 13 million curies (that amount of any radioactive substance which emits the same number of alpha rays per unit of time as one gram of radium) were released from the reactor. By April 27 the radioactive levels had fallen to about 4 million curies, and the amounts continued to decline slowly each day until they reached about 1 million curies on April 30 and May 1. The emission gradually built up to second peak on May 5 when about 7 million curies were emitted. This second peak was caused when a large amount of materials dropping on the core caused it to grow hotter and release more radioactive isotopes. Finally, on the eleventh day after the accident, May 6, liquid nitrogen was placed under the reactor to cool it, and the radiation levels receded to nearly zero on May 7. By this time the reactor fire was under control, but the long jobs of decontamination and entombing the reactor in concrete had just begun. Years of work will be required to complete the job of recovery.

SPATIAL DISTRIBUTION   The first indication outside the Soviet Union that a nuclear accident had occurred was at 2 P.M. on Sunday, April 27, when Chernobyl's radioactive cloud crossed the Swedish border. In a routine daily radiation examination on the morning of April 28, the workers at the Forsmark nuclear power station north of Stockholm registered radioactivity ten times the normal level on shoe soles. At 4 P.M. that afternoon the Swedish radio reported "10,000 times the normal amount" of cesium-137 in the air.

Within one week after the accident the nuclear contamination had covered an amazingly large area, but the spatial pattern was complex because of shifting wind patterns. During the first day and a half the winds blew northwest, carrying the invisible radioactive cloud over the Soviet republics of Byelorussia, Latvia, and Lithuania; northern Poland; and then across the Baltic Sea to the Scandinavian countries. On the second to fourth days the wind shifted to a west and southwesterly flow, blanketing the Ukraine, southern Poland, Austria, Czechoslovakia, southern Germany, Switzerland, northern Italy, and eastern France. The winds also carried the deadly radiation south and west across Romania, Bulgaria, Greece, and Yugoslavia. By the fifth day the winds shifted once again to the northwest, bringing the fallout into central Germany, the Netherlands, and the United Kingdom. On the sixth day the winds shifted northeast, carrying the cloud toward Moscow and the central portion of the Soviet Union.

The radioactive material covered at least 20 countries and extended 1,200–1,300 miles from the point of the accident. The fallout was extremely uneven. Some parts of Europe directly under the plume received little fallout while other parts received large amounts, and the materials in the plume varied greatly in weight and half-life. The amount of deposition was influenced by rainfall, which washed particles out of the air, and local topography also played a role, concentrating radioactive materials and "hot spots" in certain places and dispersing them in others.

West German and Swedish scientists found that within a radius of 60 miles radiation varied by as much as 10 to 15 times.

SOVIET DISSEMINATION OF INFORMATION The officials of the Soviet Union did not announce the occurrence of the Chernobyl accident outside the country until Monday, April 28, at 9 P.M., nearly two days after the accident, and it would appear that no announcement was made about the incident within the country until the following day. Because an increase in radiation had been detected in Sweden, there was grave concern internationally about what had occurred—it was thought, somewhere in the Soviet Union.

The first reports were sketchy, and sometimes conflicting, but clearly the Soviet leaders were anxious to demonstrate that the situation was under control. At the traditional May Day celebration, there was no indication that any emergency existed, but on May 6 *Pravda* published a detailed account of the accident, and soon the full extent of the disaster became apparent to the Soviet people and the world. In late August, the International Atomic Energy Agency (IAEA) met in Vienna, and at the meeting, the Soviets distributed a three-volume report about the accident, which said it was the result of human error.

EFFECTS OF THE CHERNOBYL ACCIDENT The effects of the accident must be viewed from two aspects; that is, from within the Soviet Union and beyond the Soviet borders. Within the Soviet Union, the effects of the accident were felt immediately at the plant site. Two workers were killed in the initial explosion, and by at least April 29 the residents of the nuclear plant city, Pripyat, and three neighboring villages had been evacuated—somewhere between 25,000 and 40,000 people. Altogether, about 135,000 people were moved permanently from the area, and a new town is being built to resettle the displaced population.

On May 1 the Soviet government reported that another 18 people were in grave medical condition owing to the accident, and on May 8 the Communist Party of the Soviet Union (CPSU) published a decree "Covering Payment and the Provision of Material Benefits to the Workers of Enterprises and Organizations in the Zone of the Chernobyl Atomic Power Station." The decree related to the job placement of the evacuees and their rate of pay and further stated the rate of pay for workers involved in the "removal of the consequences of the accident."

Two reports indicated that conditions at the plant were extremely hazardous, and there is evidence that there was much confusion in the Soviet Union as to how to handle the accident. The evacuees were frequently treated with callous indifference, and in many ways the true nature of the calamity was not revealed. For example, life went on in Kiev as if nothing had occurred at Chernobyl.

Besides the immediate damage to the plant, a considerable area of agricultural land and forests was contaminated. One hundred fifty thousand square meters of plastic film were laid on the ground in the vicinity of the plant to preserve vegetation and soil from contamination, and large areas of forest were destroyed and topsoil stripped off to reduce radiation hazards. But the soil in the region is the type that will retain radioactive particles for a long time, and the food chain will be contaminated for years.

A person can be injured by direct exposure to radioactive materials, by inhaling radioactive gases or particles, or by consuming contaminated food or water—all of which occurred in the Chernobyl area. As a result of direct exposure there were several hundred immediate deaths. Dr. Robert P. Gale, from the United States, worked with Soviet doctors to perform bone-marrow transplants on the most severely irradiated victims, but more than half of those treated have died.

The long-term health dangers from the fallout are difficult to predict, and scientific estimates of the probable number of cases of cancer and resulting deaths vary greatly. The threat from iodine-131 is difficult to estimate because it disappeared before it could be measured, and even greater uncertainty exists concerning the impact of low-level radiation at some future date. Estimates now indicate that somewhere between 5,000 and 100,000 fatalities will result from the Chernobyl accident.

The health danger to the vast Soviet and European populations that were subjected to the radioactive emissions of the Chernobyl accident is still not clear. Scientists have estimated that between 3 and 4 percent of the radioactive isotopes in the Chernobyl core were released into the atmosphere, from 50 to 100 million curies of radioactive material, including at least 50 different types of isotopes with a half-life ranging from a few months to 24,000 years. Cesium is of the greatest concern, for scientists estimate that more cesium entered the environment from Chernobyl than from all atomic bombs tested in the atmosphere.

Beyond Soviet borders, the response to the increased measurements of radioactivity varied considerably, especially in Europe. Many countries protested the long delay by the Soviets in announcing the magnitude of the accident, and the Swedish nuclear power inspector stated, "As a precaution, we would have kept our high-risk groups—children and pregnant women—indoors until the cloud passed." Indoors they would have been shielded from potent but short-lived radioisotopes such as iodine.

By the middle of May more than 20 countries had imposed restrictions on the consumption of fresh vegetables because the levels of radioactivity were above the recommended limits set by health authorities. In addition cattle grazing on contaminated grass were producing milk with radiation above acceptable levels. Governments were generally not prepared to deal with this emerging problem of contamination, and as a result, the restrictions varied greatly. In West Germany the local government of Konstanz in the southern portion of the nation severely restricted milk and vegetable consumption; in contrast, the adjacent Swiss canton, Thurgau, imposed few restrictions. Sweden developed a national program that discouraged the consumption of fresh vegetables, berries, and freshwater fish. Most of Sweden's dairy cattle were kept in barns until the pastures were tested and declared free of contamination.

In Lapland, in northern Scandinavia, the reindeer pastures were highly polluted because of heavy rainfall while the Chernobyl cloud passed overhead. About 97 percent of the reindeer slaughtered for meat in the late summer had levels of radioactivity above those recommended for human consumption. Norway and Sweden initially banned the use of this meat, but later relaxed the standards somewhat. Because cesium-137 has a long half-life of 30 years, the reindeer that graze in the radioactive area will be contaminated for many years, and consumption of reindeer meat could have a cumulative and disastrous health effect. In the eight months following the accident the normal slaughter of reindeer should have been 60,000, and the fact that the meat couldn't be sold cost the local economy $150 million in this short period. Chernobyl has been an ecological disaster for Lapland, one that endangers the entire culture of the Sami (Lapp) people.

Most countries were unprepared to cope with the magnitude of the Chernobyl disaster. In France it was initially reported that there was no radioactive fallout because of air currents moving around the country. However, independent observers reported a major rise in the levels of radioactivity, and the government finally admitted there had been a mistake in the observations. In Italy the national government did not act but left the measurement of radioactivity to local government and citizen groups. In the United Kingdom the National Radiological Protection Board attempted to minimize possible health effects. Similarly, the international agencies did little to provide standards of safety. The International Atomic Energy Agency made no recommendations concerning food or health, and the World Health Organization provided only a broad warning. The European Economic Community placed restrictions on importing fresh food from Eastern Europe for three weeks but had no scientific information on which to base judgments.

Nuclear Energy

The development of nuclear power was spectacular in the 1970s and early 1980s. Between 1970 and 1985 the number of reactors in the world increased from 66 to 372, and the installed net capacity of megawatts rose from 15,471 to 254,178. In December 1985 an additional 253,119-MW capacity was planned or under construction. After the massive oil price increases in 1973 and 1979, the industrial nations tried to decrease their dependence on imported oil, and with the shortage of indigenous fuel resources in many industrial nations, developing nuclear power to provide electricity appeared a desirable alternative. In the 1980s the growth has, however, slowed greatly.

In 1986 nuclear energy accounted for about 4 percent of the world's primary energy. About 93 percent of the installed capacity is concentrated in the industrial nations, with about 36.5 percent in Europe, 35.2 percent in North America, 11.5 percent in the Soviet Union, and 9.2 percent in Japan. The United States has the largest number of reactors, 95 with an installed capacity of 79,529 megawatts, followed by the Soviet Union, with 48 and 29,369 megawatts; France, 43 and 37,813 megawatts; the United Kingdom, 38 and 10,670 megawatts; and Japan, 33 and 29,745 megawatts.

The dependence on nuclear plants to generate electricity varies greatly from nation to nation. France has the greatest dependence because in 1973 the Messner-Pompidou government announced that it was embarking on a program to supply by means of nuclear power 70 percent of the electricity and 30 percent of the total energy needs of the nation by 1985—at that time nuclear power was considered the cheapest and best alternative to imported oil. In 1986 France secured 64.8 percent of its electricity from nuclear power. Nuclear plants are widely scattered in the country, and until recently opposition has been local and regional, focusing on particular sites rather than the total program. Other countries with a major dependence upon nuclear power for electricity include Belgium, 59.8 percent; Sweden, 42.0; Finland, 38.2; Switzerland, 34.3; Bulgaria, 31.6; West Germany, 30.0; Japan 23.0; the United Kingdom, 19.3; the United States 16.0; and the Soviet Union, 11.0 percent. Nuclear reactors have been built in a few Third World countries such as India, Pakistan, South Korea, Brazil, and Argentina, but dependence on them for a nation's electricity remains small.

Environmental Impact of the Nuclear Fuel Cycle

The environmental impact of the nuclear energy industry has been of major importance since its origin. In the processes, from initial mining to the final waste product disposition, there is the potential for radioactive emissions at a number of the steps.

TAILINGS AND MILLING   At the mining sites there are normally mill tailings, and because of their sheer volume, the Environmental Protection Agency believes that mill tailings may present "the greatest environmental impact of all waste forms over the years. In many of the tailings in western United States radioactive contaminants have been found that are above approved maximum levels for health." In the early days of mining the environmental dangers from these tailings were not recognized. They were frequently used to provide a base for roads; new home developments, parking lots, and shopping centers were built on uranium tailings; and in dry regions tailing piles were rarely stabilized to prevent wind erosion. In the early 1980s Utah Governor Scott Matheson indicated that a 23-million-ton pile of tailings in the Salt Lake City area is "the largest microwave oven in the west." These tailing piles emit radioactive materials into the atmosphere.

CONVERSION   In the conversion process uranium ore is converted and prepared into semirefined uranium oxide (U_3O_8). This oxide, known as yellowcake, is refined into uranium fluoride (UF_6), and in the process there are small radioactive waste and air emissions. The EPA has indicated that conversion plant emissions produce an average dose of much less than 1 millirem per year to individuals within 50 miles of the plant.

ENRICHMENT The next step in the nuclear cycle is the enrichment of the UF_6 by a gaseous diffusion process, which requires large quantities of water, heat, and electricity. The large quantity of cooling water required is the major environmental impact; the other environmental impacts are minimal.

TRANSPORTATION A massive amount of material must be transported from one place to another in the nuclear fuel cycle. It has been estimated that to operate a single nuclear facility of 1,000-MW capacity there are truck shipments of uranium ore totaling 16,800 miles on private land, shipments of uranium oxide and hexofluoride totaling 33,500 miles on public highways, and 2,000 miles of rail shipment of fission products. There is always the danger of an accident in transferring nuclear materials even though especially designed containers have been developed. There have been no serious accidents, but a few minor accidents have released radioactive materials into the atmosphere.

FUEL FABRICATION As enriched UF_6 is converted to uranium oxide (UO_2) and loaded into fuel cells, small emissions do occur. The EPA has indicated that an individual working at a fabrication plant might receive a maximum of 10 millirems per year in the lungs from normal breathing but the average dose within 50 miles of the plant would be less than 0.1 millirem per year.

POWER PRODUCTION The most controversial part of the nuclear fuel cycle has always been power production, and the Three Mile Island and Chernobyl accidents have intensified this controversy. The power plant is the place where new radioactive materials are produced. The fissioning of the uranium and the neutron activation produce many radioisotopes. These isotopes, particularly the gaseous ones, can escape from the fuel and through pinhole defects to contaminate the coolant water. Technical specifications limit the amounts and rates of release of these radioactive materials into the environment, and the EPA has estimated that the health effect of one year's operation of a 1,000-MW plant would be responsible for a total of only about 0.001 millirems per year. It is thus evident that the environmental contamination by radioisotopes from an operating plant is extremely minimal.

A far more significant problem relates to the cooling system. If the coolant were to stop passing through the reactor core, the fission reaction would stop, for the moderating influence of the coolant would end. At the same time the heat from the radioactive decay of the fission products would result in a melting of the fuel and if continued would lead to a break in the pressure vessel and containment system. At this point the radioactive gases would escape into the atmosphere. This type of accident could have occurred at Three Mile Island but was prevented; it did occur, however, at the Chernobyl plant in the Soviet Union.

FUEL REPR

. . . they are spent. Because they are highly radioactive, the normal method is to store them in water for a few months to reduce the radioactivity. The rods are then transported to a reprocessing plant where the wastes are removed and the usable uranium and plutonium are recovered. The EPA has attempted to predict the health effects of the reprocessing plant, and has estimated that a five-metric-ton-per-day plant, which would serve 45 nuclear plants of 1,000-MW capacity, would emit radioactive gases at a rate that would produce about 2.5 health effects annually in the United States and about 100 health effects annually worldwide.

NUCLEAR WASTE MANAGEMENT   The problem of disposing of radioactive wastes produced by nuclear power plants has attracted much attention, and public fears have become so pervasive that the industry has been threatened. The annual waste material from a 1,000-MW nuclear reactor is about two cubic meters containing about 35 kilograms of radioactive waste. Between 1970 and 1985 the metric tons of radioactive wastes increased from 6,219 to 59,600 in North America and Europe—no estimates are available for the Soviet Union or Third World countries. Of the 59,600 metric tons, 59.7 percent is located in Europe and 31.8 percent in the United States. The simplest and most obvious way to dispose of those high-level radioactive wastes is to bury them permanently, deep underground. Because of the radioactivity of the waste, special handling is necessary. It has been estimated that the United States will require about 400 1,000-MW nuclear plants to supply its electric power needs, which would mean that about half a square kilometer of space would be needed annually to store radioactive wastes. The permanent storage sites must be in areas that are free from earthquakes and volcanoes and have no possibility of contamination of the ground.

It is now thought that salt mines could provide the best burial sites.

Permanent nuclear waste disposal sites have not been chosen anywhere in the world. All waste remains temporarily stored, because whenever a site is proposed there is massive opposition. The public fears that the waste canisters will ultimately leak and contaminate the surrounding area, and there is also fear that social and political conditions are not sufficiently stable to guarantee safety. Because waste is accumulating, permanent sites must be chosen for disposal in the 1990s. In the meantime there will be much debate, and the governments may have to ultimately decide how and where to dispose of the radioactive waste materials.

Future of Nuclear Energy

In the 1970s, when nuclear programs were expanding rapidly, many governments had lavish plans for growth. Since the accident at Three Mile Island in 1979 and particularly since the Chernobyl accident in 1986, nuclear scientists have been wrestling with the implications of the safety of reactors, but opinions vary greatly. One opinion is that a disaster is always possible; another is that nuclear energy is safe and the only reliable source of energy in the future.

The overwhelming majority of nuclear scientists have stated that the probability of the type of accident on the Chernobyl scale occurring in a Western reactor is too small to be calculated. Until the Chernobyl accident no one had ever been killed, and only a few injured, by radiation from a nuclear power plant. Nuclear energy can be characterized as a low-risk but a high-dread industry. There is fear because the unknown factors cannot be calculated by the general public. The question of the development of nuclear energy is no longer one of only technology but one of human perception. Even nuclear scientists cannot definitely conclude that nuclear energy is completely safe. Thus it becomes a philosophical question, How safe is safe?

The early expectation that nuclear energy would replace all traditional sources of energy is now only a dream. Even well before the two recent catastrophes the goals had been lowered. In the mid-1970s the International Atomic Energy Agency projected that 4.45 million megawatts of nuclear energy would be in place by the year 2000. In 1986 this projection was greatly reduced to 372,000 megawatts of nuclear power by 1990 and only 505,000 megawatts by the year 2000. Despite the slowdown the nuclear power industry is large and continues to expand slowly. The reasons for reducing future growth are diverse and vary from country to country, but the major factors are the high costs, slowing electricity demand, technical problems, mismanagement, and political opposition. These factors form a complex web of interrelated problems.

In the United States, the forerunner in the development of the early nuclear power industry, the industry is now in a stage of

been withdrawn from the program. In 1985 and 1986, 14 plants were completed, and 19 more are to be completed from 1987 to 1989. After 1989, only 3 nuclear reactors are scheduled, and it is highly likely these will be canceled. Economic forces have been most effective in contributing to the decline of the nuclear power industry of the United States. The cost per kilowatt hour of capacity rose from less than $200 in the 1970s to over $3,500 in 1987. Nuclear power is in a state of transition. Its future remains a question mark.

Photochemical Air Pollution

Photochemical air pollution, commonly known as smog, has become important since the middle 1940s primarily because of the increase in automobile emissions. This type of pollution is the result of a number of complex chemical reactions. The hydrocarbons, hydrocarbon derivatives, and nitric oxides emitted from such sources as automobiles are the raw materials for photochemical reactions. In the presence of oxygen and sunlight, the nitric oxides (NO_X) combine with organic compounds, such as the hydrocarbons from unburned gasoline, to produce a whitish haze, sometimes tinged with a yellow-brown color. In the process a large number of new hydrocarbons and oxyhydrocarbons are produced. These secondary hydrocarbon products may compose as much as 95 percent of the total organics in a severe smog episode. Photochemical reactions may also increase the levels of primary pollutants such as formaldehyde and other aldehydes, which are capable of contributing to oxidant formation.

The term *oxidant* refers to chemicals that oxidize substances not easily oxidized by oxygen. Photochemical oxidants consist primarily of ozone (O_3) and nitrogen dioxide (NO_2) as well as small quantities of peroxyacetyl (PAN) and other peroxy compounds. The photolysis of NO_2 alone cannot produce the elevated levels of O_3 found in urban areas. Heavy concentrations of ozone can be produced only when hydrocarbons, aldehydes, and other reactive gases are converted to peroxy radicals by photochemical processes. Peroxy radicals change the equilibrium of the photochemical process increasing the ozone levels because the nitrogen oxide (NO) is reduced in quantity and thus little is available to react with O_3 to maintain an equilibrium. In polluted atmospheres, therefore, ozone concentrations are controlled by the ratio of NO_2 to NO, the intensity of the sunlight, and the presence of reactive hydrocarbons and other pollutants such as aldehydes and carbon monoxide (CO), which can react photochemically to produce peroxy radicals. Peroxy radicals reacting with NO cause the ratio of NO_2 to NO to increase, with a concomitant rise in O_3 levels.

Of these oxidants, ozone is the most important. This bluish gas, which is 1.6 times as heavy as oxygen, is normally found at elevated levels in the stratosphere. At the surface this "anthropogenically derived" ozone may reach peak levels as high as 0.60 parts per million. Average one-hour concentrations in the range of 0.20–0.30 parts per million are common during the summer months in southern California, and in the eastern United States the one-hour summer month concentrations can be in the range of 0.05–0.15 parts per million. Ozone concentrations reach their highest levels at midday to early afternoon during the summer months; in urban areas, ozone levels decrease at night to reach background levels. The ozone levels are reduced by chemical reactions and by contact with the earth's surface. The reaction of ozone with NO and hydrocarbons is the most important for its removal from the atmosphere. The ozone may exist from a few hours up to seven or eight hours depending upon the level of NO and hydrocarbons in the atmosphere and its absorption into the surface of the ground. On a global scale the absorption and/or destruction of O_3 on ground surfaces represents the most important means of removing O_3 from the atmosphere.

Geographic Distribution of Smog

Smog development has become a worldwide phenomenon as essentially every urban area has a smog problem owing to emissions of pollutants from motor vehicles and other sources. The model area in the world for photochemical smog is the Los Angeles basin. An important feature of the weather pattern of this area is a mass of stable high-pressure air that persists for a considerable time over the region. The basin is surrounded by mountains, which restrict air movement, and off the California coast is a relatively cold ocean current that is particularly well developed in spring and early summer. Warmer air from the Pacific Ocean is cooled as it moves over the cold water on it ~~~~~~~ ~~~~~~~~~~ ~~~~ ~~~~ ~~~~~ ~~~ ~~~~ ~~~~~~ ~~~~~~~~ ~~~ lead to strong and persistent temperature inversions over Los Angeles. Trapped surface air is prevented from rising over the mountain barriers, and the smog that develops in the Los Angeles basin reaches dangerous levels several times a year.

With the implementation of motor-vehicle emission controls, a decrease in oxidant levels (mostly ozone) was reported in 1960s and 1970s. This trend appears to have been reversed as the number of motor vehicles has increased, and the Los Angeles basin has again experienced high oxidants and severe smog conditions in recent years.

Urban versus Rural Distribution of Ozone

The development of ozone as a photochemical oxidant occurs primarily in urban areas. As mentioned, in urban areas it has a daily cycle with increases during the daylight hours and rapid decreases during the night periods. In contrast, the ozone in rural areas may persist at higher altitudes with little variation over a 24-hour period. The nighttime ozone is probably associated with nocturnal inversions produced near the ground. These inversions reduce the mixing with the NO and hydrocarbons and reduce the contact with the surface. Above the inversion layer the half-life of ozone may be as long as 80 hours.

Because ozone persists for long periods in rural areas it may be transported great distances. The high ozone level in rural areas east of the Mississippi River indicates that it has been transported from distant urban areas. Although it was once thought that photochemical oxidants were an atmospheric pollution problem of urban areas only, it must now be recognized as a regional problem. Long-range transportation makes the problem of abatement a particularly difficult one because ozone may be carried by wind currents great distances from its area of origin.

Indoor Pollution

In recent years most of the pollution publicity has been concerned with the quality of the outside air, but numerous investigations have revealed that the air inside buildings is contaminated with a large variety of pollutants as well. They range from natural contaminants, such as radon, to contaminants that are the result of human activities, smoke, and combustion by-products.

Natural Pollution

Radon is a naturally occurring gas produced by the radioactive decay of the element radium. Since radium is found throughout the earth's crust, radon is found virtually everywhere. The radon isotope of most concern, radon-222, arises from radium-226 and has a half-life of 3.8 days. The half-life of a radionuclide is the period during which any given atom has a 50 percent probability of decaying to the next member of the decay sequence. Radon itself decays and produces a series of radioactive products called radon progeny, or daughters. These are short-lived, existing no longer than 30 minutes.

Outdoor air concentrations of radon are so small that the possibility of their being a health hazard is extremely small, but indoor concentrations are typically two to more than ten times higher. Radon in houses arises primarily from radon that migrates from the soil under the house or in some cases from materials that make up the house. The distribution of indoor concentrations across the United States is broad, varying from one-tenth of typical levels to a factor of 100 times higher. The indoor concentrations are controlled by balancing between the rate of entry from sources and the rate of removal by ventilation.

The radon content of regions varies considerably throughout the United States and Canada depending upon the uranium level in the soil and bedrock. The EPA has estimated that the average soil contains about 1 part per million of uranium. Phosphate rock contains 50–125 parts per million, and granite rock contains as much as 50 parts per million in the Northeast and West. In areas of high soil gas mobility, it is generally recommended that radon-resistant housing measures be taken.

It has been determined that the entrance of radon into a building depends upon the radon production rate in the soil, meteorological factors, soil permeability, the type of building substructure, and the stack effect. The stack effect is the result of lower pressure inside a building, which means that radon moves from the earth to the lower pressure. The stack effect is a major contributor to air infiltration and can influence the release of radon in soil adjacent to a building and thus cause higher radon entry rates. Wind may also remove radon from the soil, not because of the net inflow of soil gas, but because it causes a greater exchange of air between the building and the soil

Because ~~~~ ~~~~ ~~~~ ~~~~ ~~~~ that a potentially large number of houses, particularly those recently built and those reinsulated to conserve energy, have high indoor radon levels. Measurements in homes where concentrations are in order of magnitude greater than the average suggest an individual risk of lung cancer exceeding 1 percent.

A study at the University of California at Berkeley showed that radon-exposed individuals who smoked had a much higher cancer rate than nonsmokers. It was estimated that of 1,000 male nonsmokers exposed to radon, 16 would die of lung cancer, 5 more than in a nonexposed group. Among 1,000 female nonsmokers radon exposure increases the rate from 6 to 9. The statistics are much grimmer for smokers. Of 1,000 males exposed to excess radon 172 will die of lung cancer, 49 more than among nonexposed male smokers. For women smokers radon raises the number from 60 to about 85. Why radon is more lethal for men than for women is unknown, and why smokers are at greater risk is also unknown. Some experts believe that smoke-damaged lungs will trap the radioactive radon particles more effectively.

The main health risk associated with radon comes primarily from the inhalation of the short-lived radon daughters and secondarily from the breathing of radon itself. The radon daughters are most dangerous for they are short-lived and most likely will decay at the place of original deposition in the respiratory system, which is particularly sensitive to the carcinogenic effects of radiation. Other environmental factors such as exposure to mineral dust and chemical aerosols may increase the damaging of surface cells, thus increasing the sensitivity to radiation.

Exposure to ionizing radiation can have various deleterious effects on human health. Acute exposure can damage organisms and affect the gastrointestinal and/or nervous systems to result in internal bleeding, cardiac distress, and possibly death. Exposure to low doses of radon may produce delayed effects as cells are transformed. Mutations can be deleterious in many ways by reducing resistance to many diseases, including cancer.

Carcinogenesis is thought to be a multistage process, the two principal stages being, first, the development of a tumor and, second, the growth of the tumor. Although the changes induced in cells by radiation may be irreversible, the damage may not be recognized until a tumor affects the cells. The time lapse between initiation of change and actual tumor growth can be more than 20 years, and the precise molecular mechanism that transforms healthy cells into cancerous cells is still poorly understood.

Man-made Pollutants

Combustion By-products

Combustion by-products have plagued humankind from the earliest civilizations. In primitive societies the indoor fires in the middle of a room with no chimney saturated the area with smoke. In advanced societies this type of smoke problem has been eliminated but replaced by other problems. Modern indoor combustion problems come from a number of sources.

The gas stoves that burn natural gas, found in millions of homes, are a major source of carbon dioxide (CO_2), carbon monoxide (CO), and nitric oxide (NO_X). Besides these gases there may also be emissions of aldehydes, a variety of organic gases, and particulates. The amount of pollutants depends upon burning conditions. For example, when a flame is covered by a pan there is a significant increase in the level of CO produced as well as unburned particles. Gases generated in the kitchen are rapidly spread throughout the house.

Unvented space heaters, such as kerosene and gas heaters, are major sources of CO, CO_2, NO, NO_2, and SO_2, and over 10 million space heaters are used in the United States because they reduce heating costs. Although asphyxiation is no longer a major hazard in using space heaters, they are still a major source of pollutants. The emissions depend on the type of heater and burner design. Radiant heaters burn with a relatively cool blue flame and emit large quantities of CO_2, volatile organics, and particles. Convection heaters burn hotter and have large emissions of NO and NO_2. If a kerosene or gas heater is used in a small room the emissions may result in concentrations of gases that exceed ambient or health standards. Most kerosene heaters emit SO_2 because of the sulfur in many refined fuel oils.

Wood was the normal fuel during the colonial period of the United States, but as it became scarcer, wood was replaced by coal, natural gas, and petroleum. However, since the energy crisis of the early 1970s, wood-burning fireplaces and stoves have become very popular as residential heating alternatives. It is estimated there are now 12 million wood-burning stoves in the United States, and at least 50 percent of new homes have at least one fireplace. This situation has created problems of both ind---- and outdoor pollution. It ------ ---

----- ----- ---------- -- particulates in the nation and 40 percent of certain chemicals created by fires that are oxygen deficient, including certain carcinogens. Oxygen-starved stoves also produce carbon monoxide as well as carbon dioxide. Wood-burning stoves vary greatly in efficiency ranging from 25 to over 50 percent. The low-efficiency stoves are more likely to contaminate indoor air. Combustion by-products may be released from wood-burning stoves because of improper combustion, improper installation, negative air pressure causing downdrafts, and leaks in the stove parts and flue pipes and during ash removal.

Because of the increased number of wood-burning stoves in urban areas, the EPA in 1987 established limits for emissions from new wood-burning stoves similar to those for motor vehicles. Many wood-burning stoves will now have to be equipped with catalytic converters to reduce their smoke emissions, which will add hundreds of dollars to the cost of new stoves. The regulations cover "residential wood combustion devices," which are defined as "close-chambered, combustion air-controlled" appliances. Ordinary fireplaces are not affected because the flow of air to the fire is not controlled, but fireplace inserts are covered. There is no indication that the new federal regulations will apply to existing wood stoves.

Many cities and states have now imposed regulations of their own. In July 1986, Oregon, for example, began prohibiting the sale of wood stoves that do not meet state air pollution standards, and Colorado followed suit on January 1, 1987. In Amherst, Massachusetts, the city council passed a resolution that someone in a household must have a license before a wood fire can be built, and if a fire burns improperly, the city may fine the homeowner up to $500.

The idea that something as old-fashioned and pleasant smelling as a wood stove or open fireplace can be unhealthy has been slow to be accepted by the public, but it is now recognized as a serious problem. The burning of wood is being revolutionized with a total new design for the traditional wood burner, in which the catalytic wood converter operates only when the flue gases reach $260°-315°$ C.

It is now recognized that tobacco smoking is a major indoor air pollutant. About 29-30 percent of the adult U.S. population smokes regularly, which means that a large percentage of the population, both smokers and nonsmokers, is exposed to a large variety of gas and particulate-phase contaminants. Over 2,000 different chemical compounds have been identified in tobacco smoke. Some of the common ones are nicotine; nitroamines; polycyclic aromatic hydrocarbons CO, CO_2, and NO_X; formaldehyde; acrolein; and hydrogen cyanide. The carcinogenic potential of tobacco smoke has been associated with the particulate phase, many of the gases affect the membrane of the nose and throat, and eye irritation is largely caused by the presence of acrolein and, to a lesser extent, formaldehyde. The exposure of nonsmokers to cigarette smoke can cause changes in the carboxyhemoglobin levels, systolic blood pressure, heart rate, psychomotor functions, and pulmonary resistance. The smoke from parents' cigarettes has been associated with respiratory symptoms in children, and studies of nonsmokers exposed to the smoke of coworkers have shown a reduction in small airway functions comparable to smokers who consume one to ten cigarettes per day. It has also been demonstrated that involuntary smoking may prove a risk factor in the incidence of lung cancer. In a number of studies of homes where there were individuals who smoked as well as those who didn't, the nonsmokers had a higher risk of cancer than people who lived in homes where no one smoked.

Asbestos

For decades asbestos was considered to be an excellent insulator, but in 1973 public health and environmental protection agencies began to recognize the danger of asbestos in schools and other buildings. The EPA began regulating the use of asbestos when it was declared an air pollutant under Section 112 of the 1970 Clean Air Act, and asbestos is now considered to be a very hazardous material that can cause lung cancer, mesothelioma (cancer of the chest and abdomen), and asbestosis. It is now recognized that any exposure to asbestos involves some risk. Over time, asbestos materials may deteriorate or become physically damaged, which causes asbestos fibers to become airborne. The potential for exposure increases when one is cleaning and dusting a room, but even ordinary walking in a room can cause fibers to be airborne.

Asbestos poses a special problem to school children for exposure early in life provides a long development period for asbestos-related diseases. As a result, the EPA in 1979 provided guidance to schools in identifying asbestos exposure problems. Some schools followed the early guidelines, others did not. Finally on June 23, 1983, the EPA developed a set of rules that required school districts to (1) inspect all ~~~~~~~~~~~~~~ ~~~~~~~~~~~~~~~~~~~~~~~~~~~~~~~~~~~~~stos-related matters, ~~~~~~~ school employees and parents of the presence of friable asbestos in the school environment. The rule did not require the removal of the asbestos; it was felt that pressure from parents and teachers would provide sufficient incentive to school officials.

The health danger of asbestos rests on a number of factors, including fiber concentration, age and condition, extent of use, accessibility, level of occupant activity, potential for damage, and humidity conditions. In general, high fiber content and extensive use are the most critical factors. For example, in a school environment, where there is much use and a high potential for damage, there is a strong possibility for air contamination by asbestos particles.

Formaldehyde

In recent years there has been growing concern about the indoor contamination by formaldehyde, a widely used industrial and commercial chemical. The products that have the potential to significantly contaminate indoor air include paneling, furniture, fiberboard, hardwood plywood paneling, and urea-formaldehyde foam insulation. In wood products, formaldehyde resins are used as an interior adhesive. Urea-formaldehyde resins are chemically unstable, and when the chemical decomposes it is the unchanged formaldehyde fraction that is primarily responsible for the high indoor formaldehyde levels, particularly in mobile homes, conventional homes with particleboard subflooring, and homes insulated with urea-formaldehyde foam.

The effects of indoor formaldehyde pollution may create more health problems than the carcinogens that cause cancer. The National Research Council's Committee on Toxicology has recently reported that as much as 10 percent of the population of the United States may be hypersensitive to the irritant effects of formaldehyde. To the present only a few scientific studies have been conducted on the health effects of formaldehyde, but mounting evidence indicates that the effects of this irritation include insomnia; headaches; depression; memory loss; irritation of the eyes, throat, and sinuses; dizziness; fatigue; diarrhea; and rashes.

There have been few studies of what level of formaldehyde is harmful to human health, but most homes have measurable formaldehyde levels. In an ordinary house the levels usually vary from 0.02 to 0.07 parts per million, and in one-third of the houses in the United States, which are insulated with urea-formaldehyde foam, the level may reach 0.13 parts per million. No precise health standards have been set for formaldehyde, but the National Institute for Occupational Safety and Health has indicated that the formaldehyde level should not exceed 1 part per million. This standard, however, is inadequate for it is established to protect nominally healthy workers aged 18–65 for an eight-hour day and a five-day work week. Exposed individuals in a residential environment will be exposed for many more hours each day on a continuing basis.

A number of Western European countries, including Denmark, the Netherlands, and West Germany, have proposed or established indoor air quality standards for formaldehyde, approximately 0.1 part per million maximum concentration. In the United States the American Society of Heating, Refrigeration, and Air-Conditioning Engineers has recommended that the formaldehyde level be no higher than the European standard, but this is only a recommendation, and no controls have been implemented. A great deal more scientific work needs to be done before health standards can be established, but preliminary investigations indicate that toxic irritation begins at levels more than 0.1 parts per million. There is also little toxicological information currently available to determine the significance of long-term, low-level exposure to formaldehyde.

Pesticides

It is estimated that over 90 percent of U.S. households use some type of pesticide, the most common being insecticides and antimicrobial agents, and a wide variety of chemicals are used in these pesticides. Major concern is expressed about the chlordane used to control termites, and another pesticide that has evoked concern is pentachlorophenol, a wood preservative. These pesticides may have an acute effect on human health. For example, chlorinated hydrocarbons can accumulate in human fat tissue at low exposure rates, and these hydrocarbons have an unusually high percentage of compounds that are known to be potential human carcinogens.

Biological Pollutants

In every home there are a great variety of biological pollutants that may affect the health of individuals and cause infectious or contagious illnesses. Because individuals are in close contact indoors, the transmission of these diseases occur when there is a transmission of bacteria and viruses.

Sometimes an airborne disease is so virulent that a single outbreak will secure worldwide attention. An example of such a disease is the outbreak of Legionnaires' disease in 1976 in Philadelphia at an American Legion convention. For months the cause of the sudden illness was not known, but it was finally determined that it had been caused by bacteria that were found in the air ventilation system of the hotel. The bacteria had spread throughout the hotel, but their greatest concentration had been in the lobby where the delegates congregated. This pneumonia-like disease is noteworthy because of the high mortality rate of 15–20 percent.

Somewhat related to Legionnaires' disease but less virulent is an illness known as hypersensitivity pneumonitis or humidifier fever. This illness occurs as a response to the contamination of indoor air by a variety of organisms. The disease is not of an infectious nature but rather an allergic response to inhaled particles and sensitized lymphocytes. The allergic response causes an inflammation of the respiratory passages and the lungs, and the characteristic symptoms are coughing, wheezing, chills, fever, headache, and fatigue. The outbreak of the disease is usually thought to be caused by the presence of mold or bacteria in the heating and ventilation systems of a building, which carry the bacteria to every part of the building. However, the disease can occur whenever the indoor environment encourages the growth of mold and bacteria.

In addition to the special diseases caused by bacteria, allergies may also be a response to biological pollutants. Allergies are immunological reactions to airborne particles that contain antigens, and the biological materials that can cause allergies include pollens, mold spores, and excreta from animals and insects. House dust that contains bacteria has also been identified as a cause of allergies. These indoor allergy-causing pollutants can be partially controlled by air-cleaning systems.

Indoor Air Pollution Control

Once it has been determined that a problem of indoor air pollution exists, there are a number of steps that can be taken to control it. Fundamental to controlling any pollution is the removal of its source. For example, if formaldehyde levels are considered too high, the formaldehyde-emitting wood products can be removed; the contamination of air associated with the use of gas can be minimized by using electricity; and radon contamination can be reduced by improving ventilation. If a building already in use contains a contaminant such as asbestos, the only real solution is its removal. In a number of situations it may not be practical to remove the contaminant, but in these cases, treatment measures may significantly reduce pollutant emissions.

A time-honored means to reduce indoor pollution has been to increase ventilation to remove pollutants by an exchange of air. The higher the exchange rate the greater the possibility for reducing the indoor pollutants. It must be remembered, however, that this system may exchange one pollutant for another as outdoor pollutants enter the building.

There are three means for the exchange of air from indoor to outdoor. The simplest is natural ventilation. The second is mechanical ventilation involving a forced system of air movement. The third and most complete is an infiltration/exfiltration system in which the air is not only moved but cleaned at the same time by filters.

In recent years climate control systems have become important in combating indoor air pollution. To be most effective, climate control must be a year-round operation. Climate control, depending upon its operation, can reduce or increase indoor contaminants. If the air of a building is not renewed from outside sources, the contaminants will be trapped inside. On the other hand, if the climatic control system has a ventilation outlet, there will normally be a decrease in the pollutant levels.

2

Chronology

AIR POLLUTION IS NOT A SIMPLE PHENOMENON but has developed as a response to a number of disparate events. The following critical dates may aid in the understanding of an environmental hazard that has plagued humans for hundreds of years.

Historical Air Pollution Events

A.D. 61   Seneca reports that smoke fills Rome.

1306   Edward I of England issues a proclamation forbidding the use of "sea coal" in London owing to smoke concentration.

1580   Queen Elizabeth bans the use of coal for heating buildings when Parliament is in session.

1661   John Evelyn of London writes the brochure *Fumifugium, or the Inconvenience of the Aer and Smoake of London Dissipated* for King Charles II and Parliament.

1819   English Parliament appoints a committee to study smoke in the atmosphere. This initial committee was followed by many more during the nineteenth century.

1853   British Parliament gives police power to control smoke in London. The actions are minimal.

1863   Alkali Act in the United Kingdom attempts to control atmospheric contamination from chemical plants.

1880s   Many British municipalities pass regulations to curtail smoke. These regulations are ineffective.

1920s–
1930s   Surveys of atmospheric contamination are made in such cities as New York and Salt Lake.

1949   First U.S. National Air Pollution Symposium is held in Pasadena, California.

1950   First U.S. Technical Conference on Air Pollution is held in Washington, D.C.

1955   United States passes first air pollution control act.

1956   Great Britain enacts Clean Air Act.

1959   First International Air Pollution Conference is held in London.

Air Pollution Incidents

1911   Glasgow, Scotland. Smoke-fog pollutants from coal result in more than 1,000 deaths.

1930   Meuse Valley, Belgium. Stagnant air is unable to dilute industrial pollutants; over 60 deaths are attributed to smog.

1948   Donora, Pennsylvania. Industrial pollutants concentrate because of a temperature inversion. More than 6,000 people experience respiratory problems.

1952   London, England. Smoke-smog pollutants from coal cause an estimated 4,000 deaths.

1976   Milan, Italy. Chemical plant accidentally releases a cloud of highly toxic dioxin into the atmosphere.

1984   Bhopal, India. Chemical plant accident releases deadly gases. Many deaths and more than 10,000 injuries.

Volcanic Eruptions

1815   Tambora eruptions in Indonesia cause "the year without a summer" in 1816.

1883   Krakatao eruption in Indonesia affects world weather for three years.

1888   Bandai-san eruption in Honshu, Japan, lowers temperatures in 1890 and 1891.

1890   volcanic eruption in Sicily lowers temperatures for two years.

1902–1903   Mount Pelée and Santa Maria eruptions in West Indies lower temperatures.

1912   Mount Katmai eruption in Alaska appears to affect world temperatures.

1982   El Chichón eruption in Mexico may have affected climate.

Acid Precipitation

1950s   Eville Gorham finds acid in precipitation in northwestern England.

1960s   Svante Odén, father of acid rain research, integrates knowledge of limnology, agriculture, and atmospheric chemistry. Establishes a Scandinavian network of surface water chemistry and publishes research on acid rain.

1965–1970s   Canadian lakes studied to test the effects of acidification.

1972   Gene E. Likens, F. Herbert Borman, and Noye M. Johnson publish first scientific paper presenting evidence that acid rain is prevalent over the entire northeastern United States.

1970–1980s   Studies are made of acidification of streams and lakes in the United States. Other studies are carried out to determine if acid rain has caused the decline of coniferous forests in the United States and Europe.

Carbon Dioxide

1958   C. D. Keeling of the Scripps Institute of Oceanography establishes modern monitoring of the carbon dioxide content of the atmosphere.

1960s   The effect of carbon dioxide in creating the greenhouse effect is increasingly studied.

1980s   Measurement is made of the rapid increases in carbon dioxide in the atmosphere owing to the burning of fossil fuels and the world's forests.

Ozone

1974   Mario Molina and F. Sherwood Rowland produce models demonstrating that emissions of oxides of chlorine, nitrogen, bromine, and other compounds into the atmosphere could destroy the natural balance of ozone in the stratosphere.

1976   European Economic Community agrees to reduce use of chlorofluorocarbons (CFCs).

1980   Environmental Protection Agency proposes limiting CFC production to current levels. United Nations Environment Programme Governing Council recommends that national governments reduce the production and use of CFCs.

1981   United Nations establishes an ad hoc working group of legal and technical experts to develop a global framework for the protection of the ozone layer.

1985   British scientists find a major decline of ozone over Antarctica.

3/1985   Vienna Convention for the Protection of the Ozone Layer is signed by 20 countries.

9/16/1987   Twenty-four nations sign a treaty to preserve the ozone layer by reducing the production of CFCs.

Radioactive Contamination

7/16/1945	First atomic bomb is exploded in New Mexico.
1946-1958	United States conducts atomic bomb testing on Bikini and Eniwetok in the Pacific Ocean.
1957	World's first electricity generating nuclear power plant goes into operation in Shippingport, Pennsylvania.
1958	Moratorium is placed on U.S. atmospheric atomic testing in the Marshall Islands.
1978	Bikini is still contaminated by radioactivity, and all persons are removed from the island.
3/28/1979	Three Mile Island nuclear plant accident.
4/25/1986	Chernobyl nuclear power plant explosion.

U.S. Clean Air Legislation

1955	Air Pollution Control Act of 1955, Public Law 84-159, July 14, 1955
1960-1962	Air Pollution Control Act Amendments of 1960, Public Law 86-493, June 6, 1960, and Amendments of 1962, Public Law 87-761, October 9, 1962
1963	Clean Air Act of 1963, Public Law 88-206, December 1963
1965	Motor Vehicle Air Pollution Control Act of 1965, Public Law 89-272, October 20, 1965
1967	Air Quality Act of 1967, Public Law 90-148, November 21, 1967
1970	Clean Air Act Amendments of 1970, Public Law 91-604, December 31, 1970
1977	Clean Air Act Amendments of 1977, Public Law 95-95, August 7, 1977
1988	Proposed legislation 1988, Senate Bill 1691

New Source Performance Standards (NSPS)

1970   Clean Air Act requires Environmental Protection Agency (EPA) to establish categories of sources of air pollution.

12/23/1971 EPA establishes initial New Source Performance Standards (NSPS) for coal-burning plants.

1973   Sierra Club challenges 1971 standards in court but challenge is dismissed on procedural grounds.

1975-1976   New amendments are drafted by House Subcommittee on Health and the Environment to require percentage reduction of emissions. Bill is passed by the House and Senate but dies in Conference Committee.

1-5/1977   Office of Air Quality Planning and Standards of EPA announces plans to review NSPS.

11/1977   EPA expects to issue NSPS based on full scrubbing by February 1978.

11-12/1977 Opposition to full scrubbing to reduce emissions develops in the Office of Planning and Evaluation and the Office of Research and Development in the EPA.

2/1978   Georgia Power Company and the House Science and Technology Committee oppose full scrubbing.

7/14/1978 Sierra Club asks court to require EPA to propose NSPS by August 7 as decreed by the 1977 Clean Air Act. Court sets the date of September 12 and final standards in six months.

8/1978   EPA and the Department of Energy (DOE) use computer simulation to prepare models to propose alternative NSPS.

9/12/1978 EPA and DOE present alternative models in Federal Register for comment. Full scrubbing is eliminated and partial scrubbing accepted.

9/1978-4/1979 Hearings on modeling results are held.

4/1979   EPA approves partial scrubbing based on dry scrubbing for low-sulfur coals, wet scrubbing for high-sulfur (70 to 90 percent reduction).

6-7/1979   Sierra Club, Environmental Defense Fund, Kansas City Power and Light, and other organizations petition EPA to reconsider the new NSPS.

2/6/1980   EPA denies petition to change NSPS.

9/22/1980   Environmental groups request that the U.S. Court of Appeals force the EPA to change NSPS.

4/29/1981   U.S. Court of Appeals affirms EPA's setting of NSPS.

**1982-
1988**   Extensions are granted for NSPS to obtain clean air in the United States.

3

Biographical Sketches

A LARGE NUMBER OF SCIENTISTS in universities, government laboratories, and private institutions are engaged in research on various aspects of air pollution. The following short biographical sketches of such scientists give an idea of the range of work that is in progress.

Forrest C. Alley

Alley is professor of chemical engineering at Clemson University. He received his B.S. in 1951 and M.S. in 1955 from Auburn University and a Ph.D. in environmental engineering from the University of North Carolina in 1962. He is noted for his work in industrial pollution control, exhaust emission control, and atmospheric chemistry. He is coauthor with C. David Cooper of *Air Pollution Control: A Design Approach.*

Frederick Herbert Borman

An ecologist, Borman received his B.S. degree at Rutgers University in 1948 and his M.A. (1950) and Ph.D. (1952) from Duke University. He taught at Dartmouth College from 1956 to 1966, and since then, he has been Oastler Professor of Forest Ecology, School of Forestry, Yale University. He has received the George Mercer Award and is a member of the National Academy of Sciences, Fellow of the AAAS, and a member of Ecology Society of American Research. His major area of work is the function and development of forest ecosystems.

Georg M. Breuer

Breuer is a physical chemist working in the area of industrial hygiene at the National Institute of Occupation Safety and Health. He received his B.S. from the University of Missouri at Rolla in 1966 and his Ph.D. from the University of California at Irvine in 1972. His research concentrates on the analytical chemistry of air pollution. He is the author of *Air in Danger: Ecological Perspectives of the Atmosphere.*

Jack G. Calvert

Calvert is a senior scientist at the National Center for Atmospheric Research. He received his B.S. in 1944 and Ph.D. in 1949 from the University of California at Los Angeles. He has served on many national committees, including the Committee on Health Effects of Air Pollutants, National Academy of Science, and the USA-USSR Joint Committee on Atmospheric Modelling and Aerosols. His research centers on atmospheric chemistry and reaction kinetics. He, with others, published *Acid Deposition, Atmospheric Processes in Eastern North America: A Review of Current Scientific Understanding.*

Larry W. Canter

A sanitary engineer, Canter received his B.S. from Vanderbilt University in 1961, his M.S. from the University of Illinois in 1962, and his Ph.D. from the University of Texas in 1967. He is professor of civil engineering at the University of Oklahoma and author of *Acid Rain and Dry Deposition.*

Eville Gorham

Gorham is a scientist and an educator. He received his B.Sc. in biology with distinction from Dalhousie University in 1945, his M.Sc. in zoology from the same university in 1946, and his Ph.D. in botany from the University of London in 1951. He was a lecturer in botany at University College, London, from 1951 to 1954; senior science officer at the Freshwater Biology Association, Ambleside, England, 1954-1958; lecturer and assistant professor at the University of Toronto, 1958-1962; and associate professor at the University of Minnesota from 1962 to 1984 where he has been Regents Professor of Ecology and Botany since 1984. He has been a member of the Canadian International Committee on Atmospheric Chemistry and Radioactivity, 1959-1962; a member of Scientists Institute for Public Information, since 1971; visiting member of the panel to review the toxicology program of the National Academy of Science–NRC, 1974 and 1975; a member of the Committee on Medical and Biologic Effects on Environmental Pollutants, Assembly of Life Sciences, 1976–1977; and a member of the board of directors, Acid Rain Foundation, since 1982. He has contributed many articles on limnology, ecology, and biochemistry to professional journals.

Howard E. Hesketh

A professor of chemical engineering at Southern Illinois University and a consultant for air pollution control, Hesketh received his B.S. and M.S. in 1953 and Ph.D. in 1968 from Pennsylvania State University. His research has centered on air pollution control, atomization devices, and especially the Venturi wet scrubber for both particulate and gaseous air pollution removal and fluidization. He is the author of *Fine Particles in Gaseous Media, Air Pollution Control, Controlled Air Incineration, and Wet Scrubbers.*

Joel L. Horowitz

Joel L. Horowitz is an operations research analyst at the U.S. Environmental Protection Agency. He received a B.S. from Stanford University in 1962 and a Ph.D. from Cornell University in 1967. He has done research in transportation systems analysis, econometric analysis of choice behavior, and reduction of adverse environmental impacts of urban transport systems. He is author of *Air Quality Analysis for Urban Transportation Planning.*

Sherwood B. Idso

A climatologist, Idso received his B. Physics (1964), M.S. (1966), and Ph.D. (1967), from the University of Minnesota. He is a research physicist with the Agricultural Research Services of the U.S. Department of Agriculture and an adjunct professor of geography and geology at Arizona State University. He is a member of the American Meteorological Society, American Association for the Advancement of Science, Sigma Xi, and the Royal Meteorological Society. His research concentrates on global climates, and he specializes in the effects of carbon dioxide and remote sensing.

Noye M. Johnson

Johnson, a geochronologist and hydrogeologist, is professor of geology at Dartmouth College. He received his B.S. (1953) from the University of Kansas and his M.S. (1959) and Ph.D. (1962) from the University of Wisconsin. He has been awarded research grants from the National Science Foundation and the Atomic Energy Commission and is a member of the American Association for the Advancement of Science, American Geophysical Union, and Geological Society of America. His research fields include physical limnology and stratigraphy.

Gene E. Likens

An ecologist, Likens received his B.S. degree from Manchester (Indiana) College in 1957 and his Ph.D. from the University of Wisconsin in 1962. In 1979 he received an honorary D.Sc. degree from Manchester College and in 1985 an honorary D.Sc. from Rutgers University. In 1983 he became Charles A. Alexander Professor of Biological Sciences at Cornell University, and in 1984 he was appointed professor of biology at Yale University. He is also director of the Institute of Ecosystem Studies at Cornell University and vice president of the New York Botanical Garden. He has received the Conservation Award of the American Motors Corporation, 1969; the 75th Anniversary Award, U.S. Forest Service, 1980; Distinguished Achievement Award of the Biomedical and Environmental Studies, UCLA, 1982; the first G. E. Hutchinson Award for Excellence in Research, 1982; and the Regents Medal of Excellence, Board of Regents, SUNY, 1984. He was a NATO science fellow in 1969 and a Guggenheim Fellow, 1972–1973, is a Fellow of the American Association for the Advancement of Science, and is a member of many societies and associations.

Joseph M. Marchello

Marchello, a chemical engineer, received his B.S. in 1955 from the University of Illinois and his Ph.D. in 1959 from Carnegie-Mellon University. After teaching at Oklahoma State University and the University of Maryland he was chancellor of the University of Missouri at Rolla from 1978 to 1985, and since 1985 he has been president of Old Dominion University. He has been a member of the Maryland Air Quality Control Advisory Council, Maryland Power Plant Siting Committee, Air Pollution Control Association, American Institute of Chemical Engineers, American Association for the Advancement of Science, and American Chemical

Society. He is the author of *Gas Cleaning for Air Pollution Control* and *Control of Air Pollution Sources.*

Mario Molina

An atmospheric chemist, Molina received his B.S. at the National Autonoma University of Mexico in 1965 and his Ph.D. from the University of California at Berkeley in 1972. He has taught at University of Mexico, University of California at B...

...., Jet

Propulsion Laboratory, California Institute of Technology. In 1976–1978 he was a Fellow of the Alfred P. Sloan Foundation, and in 1978 he was a Camille and Henry Dreyfus Fund teacher and scholar. He received the Tyler Ecology Award, and the Society of Hispanic Professors of Engineering Award, both in 1983. He is a member of the American Chemical Society, American Physical Society, American Geophysics Union, and the Photochemical Society. His field of specialization is the chemistry of the stratosphere.

Henry C. Perkins

Perkins, an engineer, received his B.S. (1957), M.S. (1960), and Ph.D. (1963) degrees from Stanford University. Since 1964 he has been on the aerospace and mechanical engineering faculty at the University of Arizona. He has served as a visiting professor at the Danish Technological Institute, U.S. Military Academy, and Stanford University, and has received the Teetor Award of the Society of Automotive Engineers, 1975, and the Creative Teaching Award, University of Arizona, 1982. He is the author of *Air Pollution.*

Frank A. Record

Record received his A.B. from Colby College in 1938 and his S.M. (1943) and Sc.D. (1949) in meteorology from the Massachusetts Institute of Technology. His major research interests include air pollution studies, air quality management, atmospheric diffusion, and environmental management for the control of atmospheric pollution. He worked at the Massachusetts Institute of Technology from 1949 to 1980; since 1980 he has been a consultant with GCA. He is one of the authors of the 1982 edition of *Acid Rain Information Book.*

Frank Sherwood Rowland

A chemistry educator, Rowland received his A.B. from Ohio Wesleyan University in 1948 and his M.S. (1951) and his Ph.D. (1952) from the University of Chicago. He has taught at Princeton University, the University of Kansas, and since 1963 the University of California at Irvine. He has served as a member of the Ozone Commission International Association of Meteorology and Atmospheric Physics since 1980, on the Commission of Atmospheric Chemistry and Global Pollution since 1979, on the Acid Rain peer review panel, U.S. Office of Science and Technology, Office of the White House, 1982–1984, as U.S. National Committee Member on the Science Commission on Problems of the Environment since 1986, and has been an Ozone Trends Panel Member, NASA, since 1986. He was chairman of the Gordon Conference of Environmental Sciences in 1987. He was a Guggenheim Fellow in 1962 and 1974 and has received the John Wildey Award of the Rochester Institute of Technology, 1975; Distinguished Faculty Research Award, University of California at Irvine, 1976; Professional Achievement Award, University of Chicago, 1977; and the F. J. Zimmerman Award in Environmental Science, American Chemical Society, 1980. He is a member of the U.S. National Academy of Science, the American Chemical Society, and the American Academy of Arts and Sciences.

Stephen E. Schwartz

Schwartz is a physical chemist at the Brookhaven National Laboratory. He received his A.B. from Harvard University in 1963 and his Ph.D. from the University of California at Berkeley in 1968. His major areas of research include atmospheric chemistry, chemical kinetics, interpretations of measurements of trace atmospheric constituents, laboratory studies of gas- and aqueous-phase kinetics, and modeling gas-phase and heterogeneous atmospheric reactions. He is the editor of *Trace Atmospheric Constituents: Properties, Transformations, and Fates*.

John H. Seinfeld

Seinfeld, the Louis E. Nohl Professor of Chemical Engineering and executive officer of the California Institute of Technology, received his B.S. from the University of Rochester in 1964 and his Ph.D. from Princeton University in 1967. His research has been in air pollution and optimization and control of chemical systems.

He has received many awards, including the Public Service Award from the National Aeronautics and Space Administration in 1980. He is author of *Atmospheric Chemistry and Physics of Air Pollution* and *Air Pollution: Physical and Chemical Fundamentals*.

William H. Smith

A forestry and environmental studies educator, Smith ~~~~~~ ~~~ ~~~~~~~ ~~~~ University. He taught at Rutgers University from 1965 to 1966 and since 1966 has been at Yale University as associate dean of the School of Forestry and Environmental Studies from 1981 to 1983 and since 1985, the Clifton R. Musser Professor of Forest Biology. He is a member of the Society of American Foresters and the American Phytopathology Society. He is the author of *Tree Pathology* and *Air Pollution and Forest Ecosystems*.

Arthur C. Stern

From the Stevens Institute of Technology, Stern received an M.E. in 1930, an M.S. in 1933, and an honorary Dr. Eng. Degree in 1975. His professional work has concentrated on air pollution control and industrial health, and he is an emeritus professor of air hygiene in the Department of Environmental Science and Engineering, School of Public Health, at the University of North Carolina at Chapel Hill. He has been superintendent of air pollution, New York Health Department, and chief laboratory engineer and physical scientist, Division of Air Pollution, U.S. Public Health Service. In 1970 he was awarded the Richard Beatty Mellon Award by the Air Pollution Control Association, and in 1983 he received the Christopher Bartel Award of the Stevens Institute's Union of Air Pollution Prevention Association. He is a member of the National Academy of Engineering and in 1975-1976 was president of the Air Pollution Control Association. With others, he has written *Fundamentals of Air Pollution*.

Kenneth Wark, Jr.

Wark, a member of the Department of Mechanical Engineering of Purdue University, West Lafayette, received his B.S. degree in 1950 and Ph.D. in 1955 from Purdue University and an M.S. degree in 1951 from the University of Illinois. In 1962-1963 he was an NSF Science Faculty Fellow at Stanford University. His research is in air pollution control and alternative and innovative energy conversion systems. He is the coauthor, with C. F. Warner, of *Air Pollution: Its Origin and Control.*

Laws and Legislation

SINCE AIR POLLUTION LEGISLATION BEGAN EVOLVING in the United States in the late 1940s, the efforts to control air pollution to protect public health and welfare have generated political battles. It was first believed that air pollution control was the responsibility of local and state governments, and it was not until the late 1950s that it began to be recognized that air pollution was a national problem. Even then, states' rights were strongly protected in the early federal legislation, but the ineffectiveness of voluntary local and state efforts brought control of air pollution to a prominent place in the nation's environmental programs. Environmental activism increased during the 1960s and reached a peak in 1970. In Congress, Sen. Edmund Muskie (D–Maine) and Rep. Paul G. Rogers (D–Florida) led in the shaping of national legislation, and the Clean Air Act Amendments of 1970 marked the first comprehensive attempt to provide air standards for the nation.

Local and State Air Pollution Legislation

Until the 1950s air pollution was perceived to be of local origin, affecting isolated regions, and therefore local and state governments were held responsible for providing remedies. At the local level, Pittsburgh developed one of the early successful programs to control air pollution. At the state level, the first legislation against air pollution was enacted in California as a response to worsening air pollution within the state, especially in the Los Angeles basin.

Pittsburgh and Southwestern Pennsylvania

In the nineteenth century Pittsburgh became a major center of the heavy iron and steel and associated industries, and the city came to be known as the Smoky City. It became abundantly clear after World War II that Pittsburgh's image as an industrial center, a coal town, and a railroad and river town with its grit and filth did not have a future, and civic, business, and industrial leaders saw the need for a more hospitable environment. The renaissance of the city began when political forces led by Mayor David L. Lawrence and financial and industrial forces led by Richard King Mellon joined efforts. The two men had little in common except power and the knowledge of how to use it. Mellon, from an elite, patrician family, was a conservative Republican and philanthropist; Lawrence came from an Irish worker's family.

In the middle 1940s the Allegheny Conference on Community Development was established to formulate revitalization plans. After a thorough study of the problems of the city, the commission recommended, as an initial step, the revitalization of the Golden Triangle—a 59-acre tract in the heart of the city lying at the point between the Allegheny and Monongahela rivers, which had become a warren of decayed buildings, dilapidated warehouses, parking lots, and ugly railroad yards.

Although the rehabilitation of the Golden Triangle gained the attention of the average Pittsburgh resident and provided the catalyst for change, the city's revitalization could not be successful until there was smoke abatement. The smoke pollution had reached such a level that streetlights were kept on during the daylight hours, and visibility at times was barely more than 2,050–3,000 feet because soot and dirt particles were so dense that the sunlight was weakened.

The city of Pittsburgh had attempted to control smoke as early as 1895, but an ordinance passed at that time received no support and was not implemented. A second ordinance was passed in 1941, but the need for iron and steel during World War II prevented its enforcement. After World War II the coal companies and industries objected to the imposing of any pollution controls, and only when Mellon appealed to the opposing forces did the battle end. On October 1, 1946, a smoke abatement ordinance was passed by the city of Pittsburgh, and it placed restrictions on smoke emissions from industrial and commercial buildings immediately and on home furnaces beginning October 1, 1947. In May 1949 smoke abatement control was extended to all of Allegheny County, and the Allegheny Bureau of Smoke Control was appointed. In 1960 the regulations were extended to the entire Pittsburgh region. The dieselization of the railroads and the conversion to oil and natural gas for home heating essentially solved the problems of railroad and household smoke. Emission standards ~~~~~~~~~~~~~~~~ ~~~~~~~ furnaces, and regulations forbade all outside open fires. There was thus a tight control on all atmospheric pollutants.

The efforts to control pollutants have been effective. By 1955 the hours of heavy smoke in the region had been reduced by more than 90 percent, the dustfall had been reduced from 170 tons per square mile per month to about 28 tons, and it was estimated that the annual savings in laundry and cleaning bills amounted to $41 per person. But the control of air pollutants in Pittsburgh has been costly. In 1970 it was estimated that $350 million had been spent on air pollution equipment and that expenditures continued at about $16 million per year. Although not all the pollutants have been removed from the atmosphere in the Pittsburgh region, the improvements have been spectacular. This change created the spirit that "if we could clean the smoke away, if we could make the sun visible, we could do anything," and the smoke abatement possibly led to Rand McNally's *Places Rated Almanac*'s ranking of Pittsburgh as the most livable city in the United States in 1985.

California

The California air pollution control law passed in 1950 provided the framework for later federal legislation. The key to the law was the control of emissions for it was recognized that the emission of pollutants from motor vehicles was the primary cause of air pollution in many parts of the state. The legislation found "that the control and elimination of those pollutants is of prime importance for the protection and preservation of the public health and well-being and for the prevention of irritation to the senses, interference with visibility, and damage to vegetation and property."

The pollution control legislation pertaining to motor vehicles was implemented by placing emission control devices on the fuel system. The restrictions have increasingly become more stringent, and present regulations do not permit the emission of oxides of nitrogen greater than 0.7 grams per vehicle mile, and the state board, by regulation, also provides for optional standards that are even more stringent. All vehicles are tested for compliance with emissions standards, and all persons who own vehicles that violate the provisions are liable to a civil penalty not to exceed $5,000.

The state of California has also enacted nonvehicular air pollution controls. The state has been divided into basins, and each basinwide air pollution control council is mandated to adopt a basinwide air pollution control plan.

Federal Air Pollution Legislation

By the 1950s it had become apparent that air pollution was not a local or even a regional phenomenon but a national problem so that local and state regulations could not be completely effective in dealing with it. The first federal attempts to provide a framework for the control of air pollution were tentative, but the legislative and regulatory acts between 1955 and 1970 gradually began to address the problem more directly. By 1970 the philosophical basis for federal controls had been well established even though the goal of having clean air throughout the nation has been more difficult to achieve than was envisaged by the early legislation. The legislation and its implementation have been in a constant state of change in order to achieve the goal of clean air. Although fulfillment of the goal remains in the future, efforts must persist until it is achieved.

Air Pollution Control Act of 1955
(Public Law 84-159; July 14, 1955)

The Air Pollution Control Act of 1955 was the first federally enacted legislation to prevent and control air pollution at its source. This law recognized the "dangers to the public health and welfare, injury to agricultural crops and livestock, damage to and deterioration of property, and hazards to air and ground transportation" and specifically stated that it is the "policy of Congress to preserve and protect the primary responsibilities and rights of the States and local governments in controlling air pollution." The most significant advancement was the recognition that Congress would "support and aid technical research, to devise and develop methods of abating such pollution, and to provide Federal technical service and financial aid to State and local government air pollution agencies and other public or private agencies and institutions in the formulation and execution of their air pollution abatement research programs."

This act did little to control air pollution, but it did recognize that a national problem existed. It initiated the dissemination of

step in a process that evolved over the next 15 years.

Air Pollution Control Act Amendments of 1960
(Public Law 86-493; June 6, 1960)
and Amendments of 1962
(Public Law 87-761; October 9, 1962)

In 1960 an important amendment was passed by Congress to the 1955 act. Because of the growing pollution in urban areas, it was felt that this problem had to be addressed on the national level. Congress directed the surgeon general to "conduct a thorough study for the purpose of determining, with respect to the various substances discharged from the exhausts of motor vehicles, the amounts and kinds of such substances which, from the standpoint of human health, it is safe for motor vehicles to discharge into the atmosphere under the various conditions under which such vehicles may operate." The study was completed in 1962, and as a result, the 1955 act was further amended. The studies of the surgeon general were to be continued as an ongoing process.

Clean Air Act of 1963
(Public Law 88-206; December 1963)

As a result of the studies of the surgeon general, the Clean Air Act of 1963 greatly elaborated the previous legislation. Four basic purposes of the act were defined:

1. To protect the nation's resources so as to promote the public health and welfare and the productive capacity of its population
2. To initiate and accelerate a national research and development program to achieve the prevention and control of air pollution
3. To provide technical and financial assistance to state and local governments in connection with the

development and execution of their air pollution prevention and control programs

4. To encourage and assist the development and operation of regional air pollution control programs

This act also furthered the recognition that there was a need for cooperation among federal, state, regional, and local government units to control air pollution. Specifically the act provided for

1. A coordination of research, investigations, experiments, training, demonstration, and surveys
2. Financial assistance to air pollution control agencies
3. The collection and availability of the results of studies
4. The development of air quality criteria standards
5. Federal authority to reduce interstate air pollution
6. Encouragement on the part of automotive companies and the fuel industry to prevent pollution
7. Control of interstate air pollution by means of court orders and if air pollution continues the imposition of a fine

The Clean Air Act of 1963 provided for a formal process to review and assess the status of the motor-vehicle pollution problem and also initiated cooperation between government and industry. A technical committee was to be formed with representatives from the Department of Health, Education and Welfare (HEW), the automotive industry, the manufacturers of control devices, and the producers of motor fuels. This committee was to monitor the progress toward obtaining clean air and to recommend what research and development were essential.

Motor-Vehicle Air Pollution Control Act of 1965 (Public Law 89-272; October 20, 1965)

The basic purpose of the 1965 Motor-Vehicle Air Pollution Control Act was to establish automotive emission standards. The act stated, "The Secretary shall by regulation, giving appropriate consideration to technological feasibility and economic costs, prescribe as soon as practicable standards, applicable to the emissions of any kind of substance, from any class or classes of new motor vehicles or new motor vehicle engines, which in his judgment cause or contribute to, or are likely to cause or contribute to, air pollution which endangers the health or welfare of any persons."

It was recognized that there had to be a single standard for the nation, and after hearings, the then current California state emission standards for hydrocarbons and carbon monoxide were accepted as the national standard. The motor-vehicle industries protested the establishment of a national standard indicating that such a regulation imposed a burden on states with little air pollution. However, after negotiation, they agreed to meet the single standard on the 1968-model cars. The state of California

...appropriate standards for their unique air pollution problem.

Air Quality Act of 1967
(Public Law 90-148; November 21, 1967)

After the passage of the 1965 act, establishing national standards, a heated debate evolved over the controversial issue of whether state or federal rights should prevail. On January 30, 1967, President Lyndon Johnson proposed national emission standards for industries that contributed heavily to air pollution. However, the Senate Committee on Public Works believed that regional considerations were of greater significance than national objectives. Senator Muskie rejected national standards and proposed legislation in which cooperation among all levels of government was essential. With the passage of the 1967 bill national standards were abolished, but the act did provide for a two-year study on the concept of national emission standards for stationary sources of air pollution. This study was the foundation for the 1970 Clean Air Act Amendments.

The 1967 Air Quality Act provided for

1. The establishment of air quality control regions "on the basis of those conditions, including, but not limited to, climate, meteorology, and topography, which affect the interchange and diffusion of pollutants in the atmosphere." These regions were to be "based on jurisdictional boundaries, urban-industrial concentrations, and other factors including atmospheric areas necessary to provide adequate implementation of air quality standards."

2. Development of air quality standards, "applicable to the emission of any kind of substance, from any class or classes of new motor vehicles or new motor-vehicle engines, which . . . cause or contribute

to, or are likely to cause or to contribute to, air pollution which endangers the health or welfare of any persons, and such standards shall apply to such vehicles or engines whether they are designed as complete systems or incorporate other devices to prevent or control such pollution."

3. The distribution of recommended air pollution control techniques to state and local control agencies in order to be able to achieve the levels of air quality suggested in the air quality reports. In order to accomplish this goal there was the need for government-industry cooperation.

4. The establishment of a fixed timetable for local and state agencies to enact air quality standards consistent with air quality criteria. A state could establish higher standards than proposed at the national level. If a state did not act the Secretary of HEW was empowered to establish air quality standards for each air quality region. A strict timetable provided 90 days to file a letter of intent after receiving the air quality criteria and control techniques for a specific pollutant. The standard had to be adopted within 180 days after public hearings and implemented within another 180 days. If the standards were not enforced the U.S. attorney could bring suit against a state.

In spite of the 1967 law the enactment of air quality standards lagged. The National Air Pollution Control Administration, which directed the federal efforts, was understaffed, and as a result, only a small number of air quality control regions were in operation by 1970. The procedures were extremely cumbersome and complex so that programs in the process of implementation were delayed. By 1970 it was recognized by the president and Congress that new legislation was required if air pollution was to be controlled.

Clean Air Act Amendments of 1970
(Public Law 91-604; December 31, 1970)

Provisions

The Clean Air Act Amendments of 1970, or the 1970 Clean Air Act, inaugurated the modern era of air pollution control. These amendments extended the coverage of the federal program, which had as a goal the prevention, control, and abatement of air pollution from stationary and mobile sources, and the administration of the program was removed from the Department of Health, Education and Welfare and placed in the newly created Environmental Protection Agency (EPA). The primary objective was to achieve clean air throughout the nation by July 1975.

The major provisions of the act included:

1. Expanded research funds totaling $350 million to be spent over a three-~~~~~~ ~~~~~~ to be placed on (1) the means to clean fuels before combustion; (2) the development of new and synthetic fuels; (3) epidemiological studies of the effects of air pollutants on mortality and morbidity; and (4) studying the immunological, biochemical, physiological, and toxicological effects of air pollutants.
2. An assessment of the causes and effects of noise pollution, with an Office of Noise Pollution and Abatement to be established under the Environmental Protection Agency.
3. Establishment of primary (health) and secondary (welfare) ambient air quality standards.
4. Emission limitations enforceable by both state and federal governments. Enforcement to be through criminal penalties and injunctions as well as provisions for citizen suits.
5. Transportation plans must include the "land use and transportation controls" necessary to attain air quality standards.

The key issue in the establishment of ambient air quality standards concerned the SO_2 and NO_X portions of the standard. In 1971 the EPA established a uniform system known as the New Source Performance Standards (NSPS), which stipulated standards for fossil-fuel-fired utility boilers:

1. Sulfur dioxide (SO_2)
   1.2 lb./mill. Btu heat input emission limit and a three-hour stack test to show compliance
2. Particulate
   0.10 lb./mill. Btu heat input with a normal stack test for compliance
3. Nitric oxide (NO_X)
   0.7 lb./mill. Btu emission limit on coal burned

Implementation plans to meet those standards established a specified timetable. Section 107 of the act stated, "Each State shall have the primary responsibility for assuring air quality within the entire geographic area comprising such State by submitting an implementation plan for such State which will specify the manner in which national primary and secondary ambient air quality standards will be achieved and maintained within each air quality control region in such State." The existing air quality control regions (AQCRs) were maintained, and new ones could be designated. As a result, a total of 235 AQCRs had been designated by March 31, 1975, and the total grew to 247 by mid-1975. Ninety percent of the AQCRs were intrastate.

The 1970 act also provided for the control of toxic emissions from 19 specific industries for 14 selected agents. These agents were arsenic, chlorine gas, hydrogen chloride, copper, manganese, nickel, vanadium, zinc, barium, boron, chromium, selenium, pesticides, and radioactive substances. Before a new stationary source could begin operations, state or federal inspections were required to certify that these substances would not pollute the atmosphere. Further, the compliances had to be permanent for the life of the operation.

The act also defined hazardous air pollutants and stipulated that their control would apply not only to new plants but also to existing plants. These hazardous agents included lead, mercury, cadmium, and asbestos.

Implementation of SO_2 Standards

The implementation of the New Source Performance Standards for SO_2 for large fossil-fueled steam generators with a heat input of over 250 million Btu per hour operated by utilities was a critical part of the program. The uniform standard on the limits of SO_2 emissions of 1.2 lb. of SO_2 per mill. Btu of heat input was based on an engineering judgment of what constituted "the best system of emission reduction." There had been no analysis of the impact of such a standard on the nation.

Low-sulfur coal could be burned without control equipment, but if the coal could not pass the 1.2-lb. emission limit, the coal had to be cleaned. In reality this regulation established two regional sources for steam coal—high-sulfur and low-sulfur. Most western coals are low-sulfur while most coals east of the Mississippi River are high-sulfur. Under the 1971 regulations the high-sulfur coals required flue gas desulfurization systems to remove the SO_2. These systems, known as scrubbers, use wet limestone and will remove on a daily average about 190 tons of SO_2 and consume over 400 tons of limestone and thousands of gallons of water. In 1971 when the EPA established the SO_2 standards, only three scrubbers were in existence in the United States. The EPA ruled that the technology of these scrubbers was adequate to remove 70 percent of the SO_2 and could be the basis for the established standard.

To avoid the purchase of scrubbers many utilities . . .

. . . coals are more expensive than the high-sulfur coals, the total cost was less than the investment in scrubbers. This development had a devastating effect on the coal industry in Appalachia and the Midwest. By 1978, 7,000 miners in West Virginia were unemployed, and an additional 2,000 were working only part-time. The same critical employment situation occurred in other states, and the congressional representatives from the affected states mounted a strong attack on the 1971 NSPS requirements.

The environmentalists also disapproved of the new regulations from a different viewpoint. In the West, such as at Four Corners in the Southwest, huge power plants were being built that consumed low-sulfur coal but had no pollution controls. Environmentalist groups, such as the Sierra Club, were concerned about the impaired visibility of the scenic wonders of the West, such as the Grand Canyon, and also pointed to the potential health risks of power plant emissions.

In 1973 the Sierra Club and the Oljoto and Red Mesa chapters of the Navaho Nation, whose land is located in the Four Corners area, challenged the 1971 standards in the court. The petition demanded that the 1971 NSPS require 90 percent removal of potential SO_2 emissions irrespective of the type of coal burned, which would have required the use of scrubbers in plants using western coal. The case was dismissed because the plaintiffs had not exhausted all administrative remedies.

In 1976 the Sierra Club continued its endeavors to revise the NSPS on the basis that air pollution control technology had improved so that 90 percent of the SO_2 removal was practiced and thus legally required. During the period from 1971 to 1976 a strong debate evolved in which environmentalists expressed the viewpoint that scrubbers were a necessary technology for air pollution control by utilities. In opposition the utility companies opposed any across-the-board use of scrubbers.

During this period of debate a political alliance of convenience involved the environmentalists and the high-sulfur coal industry. Environmental groups sought an amendment to the 1970 Clean Air Act that would require a set percentage reduction requirement for SO_2, a percentage reduction that would require control equipment for the burning of all coals. By this means clean air would be assured in all regions. The high-sulfur coal interests—that is, coal companies and miners' unions—altered their position and supported scrubbers for an entirely different reason. The mandated use of scrubbers for all power plants in the nation would eliminate the cost advantage for western coal. If all power companies had to purchase scrubbers, regardless of the coal burned, high-sulfur coal could be burned, and an overall scrubber requirement would also eliminate regional differences in the construction and operating costs of new plants.

The power utilities continued to be opposed to the use of scrubbers at all locations. They pointed to the technological failures that persisted in the control devices; their high cost, which raised the price of electricity to consumers; and the huge sludge disposal problem. In order to alleviate the pollution problem the utilities proposed the construction of tall smokestacks in order to lessen local air pollution problems. They also agreed to control the burning of high-sulfur coals during periods of high pollution owing to such factors as unfavorable weather conditions.

In 1975 and 1976 the House of Representatives Subcommittee on Health and the Environment proposed amendments to the NSPS that reflected "the degree of emission reduction achievable through the application of the best technological system of continuous emission reduction." This proposal changed the concept from "limitation" to "reduction" and called for a technological system that used scrubbers for the burning of all coal. The report of the House committee identified six major problems and solidified the alliance between environmentalists and high-sulfur coal representatives. The committee concluded that

1. The 1971 standards give a competitive advantage to states with low-sulfur coal and are disadvantageous to states with high-sulfur coal.

2. The standards do not provide for maximum practicable emission reduction using local coal, and therefore endanger future economic growth.

3. The standards do not expand the energy resources that could be burned in compliance with emission limits.
4. The standards aggravate compliance problems for existing coal-burning stationary sources that cannot retrofit and must compete with larger, new sources for low-sulfur coal.
5. The standards increase the risk of early plant shutdowns by existing plants, ~~with~~ ~~~~

~~...~~ standards operate as a disincentive to the improvement of technology.

Later in 1976 there was an attempt to pass legislation that reflected these concepts, but lobbying efforts on the part of the auto industry, which was seeking more lenient auto emission limits, the U.S. Chamber of Commerce, utility companies, and western industrial interests blocked the passage of any such legislation. In 1977, after the election of a new president—Jimmy Carter—and a new Congress, further Clean Air Act Amendments were passed by Congress.

Clean Air Act Amendments of 1977 (Public Law 95-95; August 7, 1977)

Provisions

The Clean Air Act Amendments of 1977 kept the fundamental framework of the 1970 act. Most areas of the nation had not attained the National Ambient Air Quality Standards (NAAQS) for one or more pollutants, and for these areas, the states were again mandated to develop a revised State Implementation Plan (SIP) by July 1, 1979, which was to provide for primary compliance with NAAQS by December 31, 1982. If the goals could not be attained by December 31, 1982, a second SIP would have to be submitted that would list the procedures for attaining the standards by December 31, 1987. All revised plans had to provide for "reasonable further progress" toward attainment on the basis of annual incremental reductions in emissions.

For areas that had attained the clean air requirements mandated by the NAAQS, the SIP must include a program to prevent the deterioration of air quality. These nondegradation areas were classified as Classes I, II, or III depending upon the amount of deterioration permitted in the future: Class I—areas in which no change in current air quality is permitted; Class II—areas in which moderate changes in emissions are permitted but stringent standards are desirable; and Class III—areas in which substantial industrial growth will be allowed but the increase in concentration of pollutants up to federal standards will be insignificant.

In both nonattainment and nondegradation areas, major new stationary sources could be constructed only with a permit and had to meet the new source standards of the law for concentrations of sulfur dioxide, carbon monoxide, nitric oxides, hydrocarbons, and photochemical oxidants. Given a strict interpretation of the law, the implication was that no new industrial development could occur in the nonattainment areas. The solution to this dilemma was found in the Federal Register of December 21, 1976, which provided an interpretation for all new and modified stationary sources of air pollution. This ruling, which is known as the "emission effect policy," indicates that a new source "can locate in a nonattainment area if its emissions are more than offset by concurrent emission reductions from existing sources in that area." This interpretation was "designed to ensure that the new sources of emissions will be controlled to the greatest degree possible; that more than equivalent offsetting emission reductions (emission offsets) will be obtained from existing sources; and that there will be progress toward achievement of the NAAQS." There was also a provision for the "banking" of offsets. That is, if the offsets achieved are considerably greater than the new source emissions, a portion of this excess can be "banked" for future growth.

In the control of emissions from motor vehicles, the 1977 act extended the deadlines for hydrocarbon and carbon monoxide and relaxed the standard for nitric oxides. The deadline for transportation controls was also extended, and the indirect source review was made discretionary for the states.

Implementation of SO_2 Standards

As in the 1970 act the implementation of the sulfur dioxide emission standards required the most attention. In implementing the 1977 act, Section III favored the burning of "locally available coals" (meaning high-sulfur) and a percent reduction requirement. This section gained the favor of the environmental groups, the coal industry, and the United Mine Workers, but Sen. Muskie, chief Senate sponsor of the Clean Air Act, opposed this amendment because it was primarily economic protectionist legislation for high-sulfur coal regions and not fundamental clean air legislation.

After much congressional debate, the enacted legislation stated that the NSPS must

Reflect the degree of emissions limitation and the percentage reduction achievable through the application of the best technological system of continuous emission reduction which (taking into consideration the cost of achieving such emission reduction and any non-air quality health and environmental impact and energy requirements) the Administration determines had been adequately d̲

̲ ̲ ̲ ̲ ̲ ̲ technological system" involving a percentage reduction, it is clear that Congress had stipulated that a percentage reduction and emission limits be established for all coals. The act itself did not specify the method to reduce the SO$_2$ content, such as the use of scrubbers, but it made it possible for the EPA to require scrubbing since certain high-percentage-reduction figures can be met only through the use of scrubbers.

When the act went to the Conference Committee the utility interests, led by their lobby group called Utilities Air Regulatory Group (UARG) and working closely with Sen. Pete Domenici (R–New Mexico), opposed uniform scrubbing as being too costly. The utilities lobbied for either no percentage reduction or at least a variable percentage reduction standard, one in which the percentage reduction required would decrease as the amount of sulfur in the coal declined. The conference report stated, "In establishing a national percentage reduction ... the Administration may ... set a range of pollutant reduction that reflects varying fuel characteristics." In addition the House conferees added, "Any departure from the uniform national percentage reduction requirement, however, must be accompanied by a finding that such a departure does not undermine the basic purposes of the House provisions and other provisions of the Act, such as maximizing the use of locally available fuels." This conference report, with the phrases "uniform national percentage reduction" and "range of pollutant production," resulted in uncertainty about congressional intent.

The EPA was confronted with conflicting interpretations, for the environmentalists pointed to the requirement for a percentage reduction of SO$_2$ and the utilities cited the conference report provision for a range of pollutant restrictions. In order to implement the 1977 legislation, the EPA proposed a uniform scrubbing standard. The studies conducted by the EPA indicated that the economic, energy, and technological impacts of a uniform or full scrubbing standard were not expected to be excessive. The next step was to determine the percentage of SO_2 to be removed in order to secure clean air.

Opposition began to develop almost immediately. In the EPA's Office of Planning and Evaluation concern was expressed as to the full scrubbing feature of the operation because of the cost for plants using low-sulfur coals. The Office of Research and Development also raised objections to full scrubbing because western power plants would have to install additional electrostatic precipitators or fabric filters, which would raise costs.

The Utility Air Regulatory Group opposed the EPA proposal standards on every ground. The caucus of the UARG concluded that emission levels were more important than percentage reduction of sulfur content, the percentage removal should be set on the basis of a declining percentage removal with declining sulfur content of coal and based on cost efficiency, scrubbers have not been substantially proved to operate effectively, and 90 percent SO_2 removal has not proved to be reliable, as well as raising many other similar objections. The Department of Energy (DOE) also opposed the EPA proposals reflecting the utility industries' position. A deputy secretary of DOE, John F. O'Leary, stated, "We don't have a demonstrated technology for the scrubbing of eastern coals." Members of Congress also expressed opposition to EPA's uniform scrubbing. A key opponent was Sen. Henry M. Jackson (D-Washington), chairman of the Senate Committee on Energy and Natural Resources. Jackson and other senators, mostly from the western part of the United States, felt that full scrubbing would endanger the development of the energy resources of the United States at a time of energy crisis.

In order to resolve the question, the EPA and the DOE began computer-based mathematical modeling to predict the impacts of alternative NSPS. As a result of this modeling, the EPA and the DOE presented two alternative proposals in the September 1978 Federal Register for comment. EPA Director Douglas M. Costle stated that the "Administration has not made a decision on which of the alternatives should be adopted." The full scrubbing option was presented first because "the Clean Air Act provides that new source performance standards (NSPS) apply from the date they are proposed and it would be easier for power plants that start construction during the proposed period to scale down to partial control than to scale up to full control should the final standard differ from the proposal."

The EPA proposal for SO_2 standards was:

1. 1.2 lbs./mill. Btu emission ceiling, except for three days per month.

2. An 85 percent reduction of uncontrolled sulfur emissions, averaged every 24 hours, with three days'

3. A 0.2 lb./mill. Btu emission floor; compliance with the emission floor would constitute compliance with the percentage reduction.

4. Credit would be allowed for precombustion fuel cleaning; sulfur removed in the pulverizer or in the bottom ash and fly ash would be credited toward the percentage reduction requirement.

5. Low-sulfur anthracite coal combustion would be covered.

6. So as not to discourage emerging front-end control technology, which can be capital intensive, the administrator, in consultation with the DOE, would ensure commercial demonstration permits for the first three full-scale demonstration facilities of solvent refined coal, fluidized bed combustion (atmospheric), fluidized bed combustion (pressurized), and coal liquefaction; under such permits 80 percent SO_2 removal, averaged daily, or 0.70 lb. NO_X emission limit for liquid fuel derived from bituminous coal would be required.

The DOE proposal for SO_2 standards was:

1. An 85 percent reduction of potential SO_2 emissions during each calendar month. Bypassing allowed as long as the percentage reduction is met.

2. A 0.80 lb./mill. Btu SO_2 emission floor not to be exceeded during any 24-hour period. A sliding scale percentage reduction required up to 85 percent on high-sulfur coals. Only minimum percentage reduction enforced for 24-hour periods if SO_2 is 0.8 lb. or less. Therefore, emissions must be greater than 0.8 lb. and reduction less than 85 percent to constitute a violation of percentage reduction.

3. A minimum of 33 percent reduction of potential SO_2 emissions to ensure that even very low-sulfur coals (below 0.8 lb.) could not be used untreated.

This DOE SO_2 standard proposal was known as the "sliding scale" standard or the "variable" standard. The Utility Air Regulatory Group also proposed SO_2 standards: ceiling of 1.2 lb. and a required reduction ranging from between 85 percent removal on a coal with uncontrolled emission of 8 lb. of SO_2 to 20 percent removal on coals with uncontrolled emissions of 1 lb. or less; compliance of these standards would be averaged on a 30-day basis; and there should be consideration of a 1.5-lb. emission ceiling.

The Natural Resources Defense Council and the Environmental Defense Fund opposed both the EPA and DOE plans as being inadequate. After receiving public comment, both the EPA and the DOE developed conclusions on the issues of monitoring, averaging times, and anthracite coal. It was felt that continuous monitoring was required but that the proposed 24-hour averaging periods could be changed to a rolling 30-day average. Most of the public comments opposed including anthracite in the standards as being too costly, and the EPA ultimately exempted anthracite from its regulations.

The major discussions centered on the impact of ceilings on coal producers and miners. The National Coal Association indicated that full scrubbing would have a devastating impact on the industry, and the EPA studies confirmed that conclusion. In order to resolve the problem, the EPA shifted from a full to a partial scrubbing standard. By using dry scrubbers a minimum reduction of 70 percent SO_2 was attained, and dry scrubbers could also be used effectively for coals with less than 3.0 lb. of SO_2 mill. Btu. Dry scrubbing could reduce control significantly. Although discussions continued, a new set of NSPS was announced in August 1979: for sulfur dioxide (SO_2), 1.2 lb./mill. Btu heat input emission limits, 0.6 lb./mill. Btu emission floor, 70–90 percent removal of potential SO_2 emissions depending upon sulfur content of fuel burned, rolling 30-day averaging, and continuous monitoring; for particulates, 0.03 lb./mill. Btu, 99 percent reduction of SO_2 on solid fuels, 70 percent reduction on liquid fuels, and manual stack test for compliance; for nitric oxides (NO_X), 0.5 lb./mill. Btu when burning subbituminous coal, 0.6 lb./mill. Btu when burning bituminous coal, and continuous monitoring.

These standards were a compromise between economic, energy, and environmental objectives and political interests, and with the implementation of the standards, court actions began immediately. Environmentalists were disturbed, for they felt the final regulations were less stringent than desirable. The Sierra Club, Environmental Defense Fund, the California Air Resources Board, and others petitioned the EPA in June and July of 1979 to reconsider the proposed standards. When the EPA refused to alter its position, and ... case to the U.S. Court of Appeals of the District of Columbia. On April 29, 1981, that court affirmed the EPA's position for air pollution controls on utilities' coal-fired steam generators. In the 253-page decision, the court upheld the EPA's rule as "reasonable" on both substantive and procedural grounds. It was the court's opinion that the objectives of the Clean Air Act were fulfilled by having a variable standard, the use of dry scrubbers, and the percentage reduction for sulfur dioxide and particulate matter. The court further concluded that the provisions encouraged technological advancements in pollution control and that the procedures provided adequate notice to the public, consideration of public views, and appropriate relations with Congress, the president, and other government agencies.

Transportation Controls To Achieve Clean Air

The 1970 Clean Air Act required that the states' plans for attaining air quality standards include "transportation controls." Although the EPA was initially reluctant to develop air pollution controls on motor vehicles, the District of Columbia Circuit Court and a federal district court in California ordered the agency to develop such standards. The major controversy that evolved concerned the amount of emission reduction auto manufacturers were to achieve. The Nixon administration recommended that by 1980 new cars be required to reduce their emissions by 90 percent. Senator Muskie's committee, embracing "technology forcing," moved the 90 percent deadline to 1975. By having such a standard, it was thought the nation's mobile emission problems would be solved.

In studies of the problem it immediately became evident that the proposed dates were not practical. Only a fraction of the cars on the road could meet the standards, and there were also many technical problems to be overcome. The development of an ambient air quality program involves three types of technical requirements. First, one must determine existing air quality levels to estimate the extent of the problem; second, one must determine how large a reduction in emissions is required to produce the desired reduction in ambient levels; and third, control technologies must be developed. It requires years to meet those requirements.

Besides the technical problems there are many administrative problems to resolve, many of which result from the sheer size of the motor-vehicle problem. Although there are about 20,000 major stationary sources of pollution, there are over 100 million cars and trucks. The quantity and extent of control present problems, but the nature of the requisite control strategies is even more critical. Transportation policy is the province of government agencies, not of private industry, and transportation policies are often made by a variety of government agencies, none of which is solely responsible for these policies and few are subject to traditional legal sanctions. Further, transportation planning occurs in a time framework foreign to the EPA and other government agencies.

In order to develop a technical and an administrative structure, the EPA needed the support of the public, but this support did not come. In contrast, the reaction was overwhelmingly critical because of a number of considerations. First, the EPA's initial plan was directed toward curbing what U.S. citizens considered to be a fundamental right: their freedom to live, work, and drive where and when they please. Second, since transportation and land use planning has traditionally been considered the responsibility of local governments, the EPA's entrance into these areas was considered an affront to local officials. Third, the public believed that the Clean Air Act was a government scheme to crack down on Detroit and private polluters. Since the transportation controls derived from this policy, the EPA regulations appeared, not as a product of national consent, but as the handiwork of overzealous bureaucrats.

During the 1970s the EPA attempted to implement clean air control plans by a number of procedures. One of the earliest was the Indirect Source Review (ISR), which was to be a mechanism for state and federal review of all land use decisions that could affect air quality. The ISR guidelines issued in April 1973 included not only shopping centers, highways, and parking lots but also residential, commercial, industrial, and institutional developments. The EPA soon deemed that the ISRs invaded traditional state and local authority and that the agency was not capable of predicting the air quality consequences of land use decisions. The ISRs were thus reduced in scope to allow the states to ignore the air quality of stationary sources located near the facility being reviewed. Further, the states could disregard pollution emitted by traffic that did not have that facility as its origin or final destination, and most important, the states were permitted to set thresholds for determining which indirect sources t~ ~~~~~ ~~

To salvage the program the EPA replaced the ISRs with Transportation Control Plans (TCPs). The TCP program differed from the ISR program in several specific ways. First, the TCPs were far more comprehensive and authoritative. Second, the states had to produce plans to attain national standards for carbon monoxide by 1977 at the latest. Third, TCPs required changes in existing transportation patterns, and they therefore created major enforcement problems because implementation required changing the driving patterns of millions of people. Fourth, although the ISRs were to have taken place as soon as practical, the TCPs were to be phased in between 1973 and 1977. To begin the program, the EPA prepared TCPs for 25 major metropolitan areas in 1973.

As in the past, the TCPs were immediately attacked by a number of groups who indicated that the plans were not feasible. In 1974 the EPA realized that it could not put all of the TCPs into effect at one time, and it decided on a single site, Boston, as the place to implement the program. Unfortunately, enforcing the Boston plan proved more difficult than expected. The first step was to require all firms with 50 or more employees to reduce their available parking space by 25 percent, but it was found that most of the possible 300 firms were exempt for a variety of reasons. The EPA could identify only about 25 firms the regulations applied to, and only about 8 of those responded to the EPA's letter. Further, it was discovered that the EPA did not have enough information to sustain legal claims in a court case, so not one case reached the U.S. attorney. As a consequence the EPA abandoned its attempt to force major cities to restructure their transportation systems.

The 1977 amendments to the Clean Air Act increased the power of the EPA, and states now had to submit adequate control plans by July 1, 1979, or face the loss of federal funds for air pollution control, sewage treatment facilities, and highway construction. Further, the 1977 act forbade the construction of new stationary source facilities in areas for which there was no approved mobile source control plan. However, even with these new regulations, the EPA gave transportation control measures low priority. Almost all states missed the 1979 deadline, but only two states, California and Kentucky, were penalized.

In 1982, when the 1977 regulations were to terminate, the question before Congress was whether to extend the deadlines. The TCPs had had only a most modest effect on providing clean air, and they had not produced a system that made realistic control demands on the states. The deadline for clean air was extended. In November 1987 the EPA once more called for all states to submit new pollution control plans for noncompliance areas within two years. The EPA would then take an additional year to examine the plans. The new program stipulated that areas that could show compliance in three to five years from approval would escape penalty; other areas would be prevented from constructing large new pollution sources and would have to reduce emission of carbon monoxide or the chemicals that form ozone by 3 percent per year above what federal programs called for or face withholding of federal highway and sewer aid.

On January 1, 1988, when the most recent deadline had expired, it was found that about 62 cities and rural areas were out of compliance for ozone and that 65 were violating the standard for carbon monoxide. Twenty-three cities were on both lists, and the 26 largest metropolitan areas were on one or the other. The director of the EPA indicated that about 10 cities, including New York, Los Angeles, and Denver, would have to restrict auto use if they were to meet the air pollution standards.

Clean air in cities remains a distant goal. If it is to be achieved, the life patterns of millions of Americans must be drastically altered. Whether or not this change can be achieved remains a major unanswered question.

Proposed Clean Air Act Amendments of 1987

In 1986 and 1987 the Senate of the U.S. Congress held hearings on four bills in order to prepare amendments to the Clean Air Act: the New Clean Air Act (S.300), the Acid Deposition Control Act of 1987 (S.321), the Clean Air Standards Attainment Act of 1987 (S.1351), and the Toxic Air Pollution Control Act of 1987 (S.1384). These bills were later incorporated into a single Senate bill (S. 1894), but the bill had not been passed as of July 1988.

Senate Bill 1894 (which became Senate Bill 1691 in 1988) is directed toward solving specific problems. To illustrate, it was recognized that SO_2 and NO_X emissions from stationary sources continue to present major problems in attaining clean air. The bill stipulates that

after D...

... major stationary source of emissions of sulfur dioxide shall operate for a total of no more than 3,000 hours, notwithstanding the remaining useful life of such source. In the case of any such source that attains an emission rate of 1.5 pounds of sulfur dioxide per million British thermal units of heat input or less, such source may operate for a total of no more than 10,000 hours. Any such source that attains an emission rate of 0.7 pounds of sulfur dioxide per million British thermal units of heat input or less shall have no limit on hours of operation or other restriction to operation in hours of peak demand under this section.

The proposed changes also recognize that air pollution is an international problem. The bill stipulates that the president shall institute negotiations with Canada and Mexico to control transboundary air pollution and establish an air quality monitoring network. The bill also stipulates stricter regulations for mobile sources of emission.

Clean Air and the Future

Although there is a recognition that clean air is a necessary prerequisite for the health and welfare of the nation, 40 years of local, state, and federal legislation has not solved the problem, and many complex issues remain unresolved. These include such questions as, What is the economic cost of clean air? Can the present economy be maintained and also achieve clean air? What are the social costs to secure clean air? If clean air is not achieved, what will be the health cost to the individual and the nation? How will the traditional patterns of life be changed? These are only a few of the questions that must be answered in the future if a clean environment is to be obtained.

Legislation Sources

United States Code Annotated. St. Paul, MN: West Publishing Company. Annual.

U.S. Code Congressional and Administrative News: Laws and Legislative History. St. Paul, MN: West Publishing Company. Annual.

U.S. Senate. Hearings before the Subcommittee on Air and Water Pollution of the Committee on Public Works. *Air Pollution 1970.* Pts. 1–3. 91st Cong., 2d sess. Washington, DC: Government Printing Office, 1970.

U.S. Senate. Hearing before the Subcommittee on Environmental Pollution of the Committee on Environment and Public Works. *Clean Air Act Amendments of 1977.* Pts. 1–2. 95th Cong., 1st sess. Washington, DC: Government Printing Office, 1977.

U.S. Senate. Hearings before the Subcommittee on Environmental Protection of the Committee on Environment and Public Works. *Clean Air Act Amendment of 1987.* Pts. 1–3. 100th Cong., 1st sess. Washington, DC: Government Printing Office, 1987.

United States Statutes at Large. Washington, DC: Government Printing Office. Annual congressional sessions.

West's Annotated California Code: Vehicle. St. Paul, MN: West Publishing Company. Annual.

5

Directory of Organizations

THIS CHAPTER IS DIVIDED into three major parts. The first considers the private organizations in the United States that represent a discipline, such as the American Meteorological Society, or a group of individuals who want to enhance the environment. The second part lists U.S. government agencies, of which the Environmental Protection Agency is the major environmental agency. There are many agencies that have a specific and limited objective; these agencies may be permanent organizations or may be abolished when their objectives are completed. The third section lists international organizations, which frequently have a coordinating responsibility.

Private Organizations in the United States

Acid Precipitation Digest
Center for Environmental Information
33 S. Washington Street
Rochester, NY 14608

Description: A general reference publication on acid precipitation.

Purpose: To provide sources of information for the study of acid precipitation.

Activities: Numerous sources of information identified from the world's literature through the scanning of periodicals, from manual and computer-assisted literature searches, and from correspondence with readers, books, proceedings, periodicals, technical papers, government documents, dissertation pamphlets, and audio-visual materials are highlighted in annotated and nonannotated entries. A calendar provides comprehensive listings of future conferences, seminars, and the like. The Digest provides news summaries of current interest and also reports research opportunities. The Digest encourages readers to forward all types of materials to the editor for annotation.

Publications: None

Acid Rain Foundation, Inc.
1630 Blackhawk Hills
Saint Paul, MN 33122

Description: Founded in 1981 as a publicly supported, tax exempt, nonadvocacy organization.

Purpose: To foster a greater understanding of the acid rain problem and to help bring about its resolution by developing and raising the level of public awareness, supplying educational resources and materials to a wide range of audiences, and supporting research.

Activities: Provides information for acid rain resources directory, international speakers, reference lists and library materials, and acid deposition research; maintains an expert referral service, a computerized information retrieval system, and a library; supplies information packets and curriculum materials.

Publications:

Acid Rain in Minnesota
Acid Rain in North Carolina

Acid Rain Information Clearing House (ARIC)
Center for Environmental Information, Inc.
Information Services
33 S. Washington Street
Rochester, NY 14608

Description: Founded in 1982 as a program of the nonprofit Center for Environmental Information.

Purpose: To act as a resource center for technical and general information on the topic of acid deposition.

Activities: Provides information through its special library, publications, reference services, and conferences. Provides specific bibliographic data in response to requests.

Publications:

Acid Precipitation Digest (monthly)

American Association of State Climatologists (AASC)

c/o Prof. David R. Miller
Renewable Natural Resources Department
W. B. Young Building
University of Connecticut
Storrs, CT 06268

Description: Founded in 1976. Consists of 105 members who are state-supported and university-related climatologists. Committees deal with computers, education, ...

Purpose: Promotes applied climatology and climatological services in the United States.

Publications:

State Climatology (quarterly)

American Council on the Environment (ACE)

1301 20th Street, N.W., Suite 113
Washington, DC 20036

Description: Established in 1972. Membership of 470.

Purpose: Serves business and professional people, unions, and individuals who want to enhance the environmental quality of the United States by means of a balanced approach to the economic and social well-being of the nation.

Publications:

Newsletter (irregular)

American Meteorological Society (AMS)

45 Beacon Street
Boston, MA 02108

Description: Founded in 1919. Membership of 11,000 and staff of 38. There are 81 working groups. Consists of meteorologists, oceanographers, and hydrologists.

Purpose: To develop and disseminate knowledge of the atmosphere and related oceanic and hydrospheric sciences. This is the major academic society for meteorologists in the United States.

Activities: Guidance service, scholarship program, career information, and certification of consulting meteorologists. Issues policy statements on such topics as weather modification, forecasting, tornadoes, flash floods, and hurricanes; prepares educational films; maintains a library.

Publications:

Journal of Climate and Applied Meteorology (monthly)
Journal of Physical Oceanography (monthly)
Journal of the Atmospheric Sciences (monthly)
Meteorological and Geoastrophysical Abstracts (monthly)
Newsletter (monthly)

Association for Rational Environmental Alternatives (AREA)
P.O. Box 771885
Houston, TX 77215

Description: Established in 1975. Membership of 150 consists of professionals in environmental fields dedicated to the advancement of private nongovernment alternatives for the planning and use of the rural and urban environment. Operates through councils on energy, environmental appearance, land use, and pollution.

Purpose: To investigate the impact of government actions on urban and rural development and to encourage solutions to environmental problems. Approves regulations and restrictions on private property rights.

Publications:

Bibliographies
Environmental Alternatives (quarterly)
Policy papers
Reprints

Association of Environmental and Resource Economists (AERE)
1616 P Street, N.W.
Washington, DC 20036

Description: Membership consists of about 500 professional economists and economists from universities and government agencies interested in environmental and resource problems. Annual meetings are held.

Purpose: Primarily to consider the problems and concerns in resource management, including water and land resources and air pollution.

Publications:

Journal of Environmental Economics and Management (quarterly)

Association of Environmental Scientists and Administrators (AESA)
12375 Mt. Jefferson Terrace, 6-J
Lake Oswego, OR 97034

Description: Established in 1983 to promote the goals and interests of environmental professionals. Regional meetings are held.

Purpose: Attempts to demonstrate the value of environmental science by creating a forum in which environmental issues can be debated.

Publications:

Short briefs

Citizens for a Better Environment (CBE)
33 E. Congress, Suite 523
Chicago, IL 60605

scientists and attorneys. Maintains a substantial library.

Purpose: Works to reduce exposure to toxic substances in air, water, and land through research, public information, and formal and informal interactions. Focuses on research, public information, and court hearings and files suits in state and federal courts.

Publications:

Environmental Review (bimonthly)
Fact sheets
Research reports

Earthcare Network (EN)
c/o Michael McCloskey
730 Polk Street
San Francisco, CA 94109

Description: Founded in 1981. Consists of 16 environmental groups and research centers to create an intelligence base for the education and training essential to the continued development of global environmental protection.

Purpose: To develop research and educational programs for the dissemination of scientific information on public environmental policy. Seeks modification of policies to protect the environment, such as policies concerning carbon dioxide in the atmosphere.

Publications:

National Resource Technical Bulletin (newsletter; quarterly)

Electric Power Research Institute
3412 Hillview Avenue
Palo Alto, CA 94304

Description: Established in 1972. Study groups are the Ecological Studies Program and the Environmental Physics and Chemistry Program.

Purpose: To assess the utilities' contribution to any damage that might be attributed to acid deposition and to provide the industry, and others, with methods for evaluating different policy options.

Activities: Assesses the effects of acidic deposition on ecosystems, evaluates the utilities' contribution to acidic deposition, assesses damage costs, and develops analytical decision tools.

Publications:

Guide (three times a year)
Journal (nine times a year)
Research report (monthly)

National Acid Precipitation Assessment Program (NAPAP)
Oak Ridge National Laboratory
Martin Marietta Energy Systems
P.O. Box X
Building 2001
Oak Ridge, TN 37831

Description: Established under Title VII of P.L. 96-294, the Acid Precipitation Act of 1980. Managed by the Interagency Task Force on Acid Precipitation, which functions under the Council of Environmental Quality.

Purpose: To increase public understanding of the causes and effects of acid precipitation. Through research, monitoring, and assessments, it attempts to improve the scientific basis for making decisions on acid rain.

Activities: Through its Acid Deposition Research Inventory, collects information on ongoing and planned acid deposition research projects that are funded by the federal or state governments. Project information, which is available to the general public and is computerized and searchable interactively, is not bibliographic but rather describes the research project.

Publications:

Operating Research Plan: Inventory of Research

National Weather Association (NWA)
4400 Stamp Road, Room 404
Temple Hills, MD 20748

Description: Organized in 1976. Members are individuals and groups interested in operational meteorology and daily weather forecasting. Committees include climatology, atmospheric aviation, synoptic, radar, and tropical.

Purpose: Promotes professionalism in practical meteorology. Develops solutions to problems faced by people working in daily weather forecasting. Encourages the exchange of weather data. Presents awards to forecasters.

Publications:

National Weather Digest (quarterly)
Newsletter (eight times a year)

University Corporation for Atmospheric Research (UCAR)
Box 3000
1850 Table Mesa Drive
Boulder, CO 80307

~~~~and related fields. Operates the National Center for Atmospheric Research under contract to the National Science Foundation.

Purpose: Conducts research and fosters the development of facilities and related services in the international academic community concerned with research in the atmospheric, oceanic, and related sciences.

Publications:

Annual report
Newsletter (bimonthly)

U.S. Government Agencies

Environmental Protection Agency (EPA)
401 M Street, S.W.
Washington, DC 20460

Description: Established in the executive branch as an independent agency on December 2, 1970. In addition to main office, there are ten regional offices.

Purpose: To protect and enhance the environment today and for future generations to the fullest extent possible under the laws enacted by Congress. The agency's mission is to control and abate pollution in the areas of air, water, solid waste, pesticides, radiation, and toxic substances; its mandate is to mount an integrated, coordinated attack on environmental pollution in cooperation with state and local governments.

Activities: There are five major branches. Air and Radiation Programs—air control and quality activities include development of national programs, technical policies, and regulations for an oil pollution control; enforcement of standards; development of national standards for air quality; emission standards for hazardous pollutants; technical direction, support, and evaluation of regional air activities; and provision of training in the field of air pollution control. Related activities include technical assistance to states and agencies having radiation protection programs and a national surveillance and inspection program for measuring radiation levels in the environment. Water Programs—water quality activities represent a coordinated effort to restore the nation's waters. Solid Waste and Emergency Response Programs—this office provides policy, guidance, and direction for the EPA's program to manage solid waste in the United States. Pesticides and Toxic Substance Programs—this branch is responsible for developing national strategies for the control of toxic substances; directing pesticide and toxic substance enforcement activities; developing criteria for assessing chemical substances; establishing standards for test protocols for chemicals, rules and procedures for industry reporting, and regulations for the control of substances deemed to be hazardous to humans or the environment; and additional activities for the assessment and control of toxic substances and pesticides. Research and Development—this office is responsible for a national research program in pursuit of technological controls of all forms of pollution. It directly supervises the research activities of the EPA's national laboratories and gives technical policy direction to those laboratories that support the program responsibilities of the agency's regional offices. Close coordination of the various research programs is designed to yield a synthesis of knowledge from the biological, physical, and social sciences that can be interpreted in terms of total human and environmental needs.

General functions of the agency include management of selected demonstration programs; planning for environmental quality monitoring programs; coordination of agency monitoring efforts with those of other federal agencies, the states, and other public bodies; and dissemination of agency research, development, and demonstration results.

Regional offices: The basic purpose of these offices is the development of strong, local programs for pollution abatement. Regional administrators are responsible for accomplishing within their region the national program objectives established by the agency. They develop, propose, and implement an approved regional program for comprehensive and integrated environmental protection activities.

Regional office addresses

Region I	John F. Kennedy Federal Building, Boston, MA 02203
Region II	26 Federal Plaza, New York, NY 10278
Region III	841 Chestnut St., Philadelphia , PA 19107
Region IV	345 Courtland St., N.E., Atlanta, GA 30365
Region V	230 S. Dearborn St., Chicago, IL 60604
Region VI	1201 Elm St., Dallas, TX 75270
Region VII	726 Minnesota Ave., Kansas City, KS 66101
Region VIII	999 18th St., Denver, CO 80202

Region IX   215 Fremont St., San Francisco, CA 94105
Region X   1200 6th Ave., Seattle, WA 98101

Publications:

EPA Journal

National Air Data Branch (NADB), Data Collection Laboratory

U.S. Environmental Protection Agency, Office of Air and Standards

research Triangle Park, NC 27711

Description: The NADB is part of the Data Collection Laboratory, which, in turn, is under the EPA.

Purpose: Responsible for collecting, analyzing, and disseminating data relating to the air quality of the United States.

Activities: Maintains two computerized data systems: SAROAD (Storage and Retrieval of Aerometric Data), which contains data collected by air monitoring stations located throughout the United States, and NEDS (National Emissions Data System), which contains data on pollutant emissions and their sources across the country. The machine-readable SAROAD data bank contains 400 million aerometric data items from more than 5,000 air monitoring stations dating from 1958 to the present. The machine-readable NEDS data bank lists 140,000 point sources of emissions in 3,300 areas of 55 states and territories.

Aerometric and emission data from the systems are distributed in machine-readable form to authorized persons. Data are also available from the ten regional offices. Data are made available to air quality organizations, with some restrictions on data use.

The NADB systems have been collectively referred to as the Aerometric and Emissions Reporting System (AEROS). There are plans to upgrade them into an interactive, user-oriented computerized data management system to be called Aerometric Information Reporting System (AIRS).

Publications:

Air Quality Data from the National Air Surveillance Networks and Contributing State and Local Networks (annual)
Air Quality Data Report (annual)
Emission Data Report (annual)

National Oceanic and Atmospheric Administration (NOAA)

Department of Commerce
Rockville, MD 20852

Description: Formed on October 3, 1970. Principal functions are authorized by Title 15, Chapter 9, United States Code (National Weather

Service), the Weather Modification Reporting Act of 1972, and the Land Remote Sensing Act of 1984.

Purpose: To explore, map, and chart the global ocean and its living resources and to manage, use, and conserve these resources; to describe, monitor, and predict conditions in the atmosphere, ocean, sun, and space environments; to issue warnings against impending destructive natural events; to assess the consequences of inadvertent environmental modification over several periods of time, from 1 year to 100 years; and to manage and disseminate worldwide environmental data through a system of meteorological centers.

Activities: Reports the weather of the United States and its territories and provides weather forecasts to the general public; issues warnings against such destructive natural events as hurricanes, tornadoes, floods, and tsunamis; and provides services in support of aviation and marine activities, agriculture, forestry, urban air quality control, and other weather-sensitive activities. The agency also monitors and reports all nonfederal weather modification activities conducted in the United States.

In addition, NOAA provides satellite observations from the environmental satellite system, and conducts an integrated program of research and services relating to the oceans and inland waters, the lower and upper atmosphere, and the earth to increase understanding of the geophysical environment.

Field Organizations:

National Weather Service

National Meteorological Center	5200 Auth Rd., Camp Springs, MD 20223
Eastern Region	585 Stewart Ave., Garden City, Long Island, NY 11530
Central Region	601 E. 12th St., Kansas City, MO 64106
Southern Region	819 Taylor St., Fort Worth, TX 76102
Western Region	125 S. State St., Salt Lake City, UT 84147
Alaska Region	Box 23, 701 C St., Anchorage, AK 99513
Pacific Region	Room 4110, 300 Ala Moana Blvd., Honolulu, HI 96850

National Environmental Satellite, Data and Information Service
National Climatic Data Center, Federal Bldg., Asheville, NC 28801

Office of Oceanic and Atmospheric Research
Environmental Research Laboratories, 3100 Marine St., Boulder, CO 80302

U.S. Government Advisory Committees

The U.S. government has established a number of advisory committees to assist departments and agencies. The term *advisory committee* was defined by Congress in 1972 as "Any committee, board, commission, council, conference, panel, task force, or other similar groups or any subcommittee or other subgroup thereof . . . which is (1) established by statute or reorganization plan, or (2) established or utilized by the President, or (3) established or utilized by one or more agencies, in the interest of obtaining advice or recommendations for the President or one or more agencies . . .

. . . advisory committee is to provide advice for specific tasks; at times the committee may coordinate programs.

In practice, there are presidential advisory committees and public advisory committees. A presidential advisory committee's main function is to advise the president; the public advisory committees are permanent and provide advice for regulating activities within the jurisdiction of an agency or a department.

Acid Rain Coordination Committee
Council on Environmental Quality
722 Jackson Place, N.W.
Washington, DC 20006

Description: Established August 2, 1979. An interagency committee, which functions under the sponsorship of the Council on Environmental Quality, Executive Office of the President, consisting of representatives from the Departments of Agriculture, Commerce, Energy, Interior, and State; Council on Environmental Quality; Environmental Protection Agency; National Science Foundation; and Office of Science and Technology Policy. Committee is cochaired by the Department of Agriculture and Environmental Protection Agency representatives. Committee meets periodically, no set times.

Purpose: To plan a comprehensive acid rain assessment program and to ensure that the results of the assessment program be incorporated in agency planning and decision making. The committee seeks advice from state and private research groups in order to avoid duplication of research programs and also coordinates efforts with Mexico and Canada.

Publications:

Annual report presents results of the acid rain assessment program and makes recommendations for future programs.

Advisory Committee on Atmospheric Carbon Dioxide
(environment)
U.S. Department of Energy
Washington, DC 20585

Description: Originally established by the Energy Research and Development Administration in 1976 as an ad hoc advisory committee on the global environmental effects of carbon dioxide. In 1977 it was transferred to the Department of Energy, and its name was changed in 1979. Committee consists of representatives from the fields of ecological sciences, meteorology, oceanography, and engineering. Members are selected on the basis of their scientific background and administrative experience. Committee meets when called.

Purpose: Provides guidance in the area of environmental research dealing with the global effects of increasing levels of carbon dioxide produced by fossil fuel combustion and related matters and surveys ongoing and planned national and international research activities in the field.

Clean Air Scientific Advisory Committee (air quality)
Environmental Protection Agency
401 M Street, S.W.
Washington, DC 20460

Description: Established February 9, 1978, under the authority of the Clean Air Act Amendments of 1977. Although the committee functions within the jurisdiction of the Environmental Protection Agency, it is a public advisory committee and a subcommittee of the National Science Board. The committee consists of seven members appointed by the administration of the EPA and must include one member from the National Academy of Science, one physician, and one person representing state air pollution control agencies. Committee meets three to six times a year.

Purpose: Provides independent advice on the scientific and technical aspects of issues related to the criteria for air quality standards, research related to air quality, sources of air pollution, and the strategies to attain and maintain air quality standards and to prevent significant deterioration of air quality.

Council on Environmental Quality (CEQ)
722 Jackson Place, N.W.
Washington, DC 20006

Description: Established on January 1, 1970, under the authority of the National Environmental Policy Act of 1969. The council is a presidential advisory committee under the jurisdiction of the Executive Office of the President. Consists of three members appointed by the president and approved by the Senate; staff is provided by the Office of Environmental Quality. Meetings are called by the chairman.

Purpose: Analyzes important environmental conditions and trends, reviews and appraises federal government programs having an impact on the environment, recommends policies for protecting and improving the quality of the environment, and assists in the preparation of the president's annual report to Congress.

Publication: List of publications on request

Environmental Advisory Committee
U.S. Department of Energy
Washington, DC 20585

Description: Established in December 1976 as a public advisory committee within the Energy Research and Development Administration and was transferred to the Department of Energy in 1977. Composed of 21 members from the fields of science, medicine, engineering, and industry; state and local groups; environmental organizations; and other groups from the general public. Committee meets four times a year.

Purpose: Advises the Department of Energy on matters that pertain to the research, development, and demonstration of energy technologies and policies which affect the environment and safety of the general society. The advice includes the preparation of environmental impact assessments and statements, environmental research, and the social, economic, and institutional impacts of energy activities.

Environmental Measurements Advisory Committee
(environmental quality)
Environmental Protection Agency
401 M Street, S.W.
Washington, DC 20460

Description: Established on November 3, 1975, under the jurisdiction of the Science Advisory Board of the Environmental Protection Agency. Consists of up to 15 members who are scientists and engineers from relevant disciplines with a demonstrated ability in analytical methodology and instrumentation and with well-established knowledge and experience in environmental monitoring. Committee meets quarterly.

Purpose: Provides administrators with expert and independent advice on issues related to scientific and technical problems associated with environmental measurements and monitoring. Problems include the detection, identification, quantification, and continual monitoring of biological, chemical, and physical pollutants in air, water, soil, other environmental media, and human and plant tissues. The pollutants of concern are pathogenic bacteria and viruses, pesticides and other toxic or hazardous chemicals, and radiation, noise, and solid wastes.

Federal Agency Liaison Committee for Consultation on Air Quality Criteria

Environmental Protection Agency
401 M Street, S.W.
Washington, DC 20460

Description: Established in September 1968 as an interagency committee to function within the Environmental Protection Agency. Members are from Departments of Agriculture, Commerce, Defense, Education and Welfare, Health, Housing and Urban Development, Interior, Justice, Labor, Transportation, and Treasury; U.S. Atomic Energy Commission; Federal Power Commission; General Services Administration; National Aeronautics and Space Administration; National Science Foundation; Tennessee Valley Authority; U.S. Postal Service; and the Veterans Administration. Meetings are held at the request of the chairman.

Purpose: Advises the EPA on the establishment of air quality standards and provides lists of substances that are hazardous air pollutants.

President's Air Quality Advisory Board

Environmental Protection Agency
401 M Street, S.W.
Washington, DC 20460

Description: First established on November 21, 1967, as part of Section 117 of the Clean Air Act; on December 2, 1970, the board became a public advisory committee within the Office of Air and Water Programs of the Environmental Protection Agency. Consists of 15 members appointed by the president and selected from various interstate, state, and local government agencies; public and private interests contributing to, or affected by, air pollution; and other informed individuals. Meetings are called by the chairman.

Purpose: Advises the president and administrators on matters of policy relating to the activities and functions of the Clean Air Act.

International Organizations

Canada-United States Environmental Council (CUSEC)

c/o Defenders of Wildlife
1244 19th Street, N.W.
Washington, DC 20036

Description: Founded in 1974. Consists of representatives from national and regional environmental groups.

Purpose: To facilitate the exchange of information and aid in cooperative action on current environmental issues of the two nations such as acid precipitation, Great Lakes pollution, and the future of the Antarctic and the oceans.

Activities: Coordinates information provided by the two nations on environmental problems

Commission on Air Pollution Prevention (CAPP)
Graf-Recke-Strasse 84
Postfach 1139
D-4000 Düsseldorf 1
Federal Republic of Germany

Description: Established in 1957. It has 1,200 members and staff of 23.

Purpose: To establish effective scientific and technical guidelines that address the problems of air pollution and provide a means of prevention.

Activities: The work of the commission is carried on through five committees: Effects of Dust and Gases, Formation and Reduction of Emissions, Measurement Technology, Transport and Turbulent Diffusion, Waste Gas Cleaning Processes and Dust Technology. Three to five conferences are held each year.

Publications:

   Manual on Air Pollution Prevention (in English, 30 times a year)
   Staub-Reinhaltung der Luft (in German, monthly)
   Schriftenreihe der VDE—Kommission Reinhaltung der Luft (in German, irregular)
   VDI-Berichte (in German, irregular)

European Committee of the International Ozone Association
9 Rue de Phalsbourg
F-75854 Paris Cedex 17
France

Description: Established in 1976. Consists of 160 members of the regional groups of the International Ozone Association.

Purpose: To promote research on ozone and its industrial applications and to act as liaison among academic institutions, industries, government agencies, conservation groups, other organizations interested in the ozone, and the public.

Activities: Collects and disseminates research data on ozone technology and its application. Holds workshops, conferences, and symposia; gives a biennial prize for original paper on the utilization of ozone technologies in industry.

Publications:

OSE-Ozone Science and Engineering (quarterly)
Ozonews (bimonthly)
Who Is Who in Ozone? (annual)

International Association of Meteorology and Atmospheric Physics (IAMAP)
P.O. Box 3000
Boulder, CO 80307

Description: Founded in 1919 by interested nations from the membership of the International Union of Geodesy and Geophysics.

Purpose: To promote research in all aspects of atmospheric physics and to coordinate research in those fields that require international cooperation.

Activities: Work is carried on through commissions, which include cloud physics, climate, ozone, dynamic meteorology, upper atmosphere, polar meteorology, and radiation. Meetings are held periodically.

Publications:

Handbook of IAMAP
Proceedings of meetings

International Council of Scientific Unions
51 Boulevard de Montmorency
F-75016 Paris
France

Description: Established in 1919. Consists of 71 national research councils on academics and 20 scientific unions of various scientific disciplines.

Purpose: Facilitates and coordinates the activities of international scientific unions in the exact and natural sciences. Through the national cooperating organizations it enters into relations with governments to promote scientific research. Collaborates with the United Nations and its agencies.

Activities: The work of the council is carried on through various committees, the most important being Problems of the Environment, Solar Terrestrial Physics, Oceanic Research, and Toxic Waste Disposal. General committee meetings are held annually; general assembly meets once every two years.

Publications:

ICSU: A Brief Survey
ICSU Organization and Activities
Newsletter (quarterly)
Yearbook

International Union of Air Pollution Prevention Associations

Brighton, East Sussex BN1 1RG
England

Description: Established in 1965. Membership consists of representatives from 26 national air pollution associations.

Purpose: To promote global public education relating to the importance of clean air methods and results of air pollution control. The union also acts as liaison with other international and national scientific and technical organizations.

Activities: Facilitates the exchange of information and publications, encourages the use of uniform scientific and technical terminology, and promotes uniform methods of measurement and monitoring. Conference is held every three years.

Publications:

IUAPPA Handbook (annual)
IUAPPA Newsletter (in English and French, bimonthly)
IUAPPA World Congress Proceedings (triennial)

World Meteorological Organization (WMO)
Case Postale No. 5, CH-1211
Geneva 20, Switzerland

Description: First organized in 1878 as the International Meteorological Organization (IMO); in April 1951 the functions and assets of IMO were transferred to the World Meteorological Organization (WMO) in accordance with a convention adopted in Washington in 1947. WMO, a special agency of the United Nations, has a membership of 159 countries and territories.

Purpose: To facilitate worldwide cooperation in the establishment of networks of stations for making meteorological and hydrological or other geophysical observations and to promote the establishment and maintenance of meteorological centers charged with the providing of meteorological services; to promote the establishment and maintenance of systems for the rapid exchange of weather information; to promote standardization of meteorological observations and ensure the uniform publication of observations and statistics; to further the application of meteorology to aviation, shipping, water problems, agriculture, and other human activities; and to encourage research and training in meteorology and to assist in coordinating the international aspects of such research and training.

U.S. contacts:

Bureau of International Organization Affairs
Department of State
Washington, DC 20520

National Oceanic and Atmospheric Administration
Department of Commerce
Rockville, MD 20852

Organization Sources

Encyclopedia of Associations. Vol. 4, *International Organizations.* Detroit, MI: Gale Research Company, 1987.

Encyclopedia of Associations. Vol. 1, *National Organizations of the U.S.* Detroit, MI: Gale Research Company, 1987.

United States Government Manual, 1986/87. Washington, DC: Government Printing Office, 1986.

6

Bibliography

THERE HAS BEEN A MASSIVE INCREASE in the literature on atmospheric pollution since 1970 as a response to the recognition of not only the potential danger to the physical and biological environment but also the effects on human health. The literature varies from popular accounts to the most technical scientific papers. The selection of bibliographical items in this chapter provides a wide perspective on modern atmospheric problems. The final section of the chapter consists of a listing of selected journals that publish articles on the atmospheric environment. In addition to the books, articles, and journal titles listed in this chapter, there are other publications and data banks that will be of interest to the reader. Many of these sources can be found in Chapter 4, listed under the organizations' activities and/or publications.

Books

General

Beery, W. T., ed. **Air Quality—Status, Management, and Directions: Summary of the Tenth APCA Government Affairs Seminar.** Pittsburgh, PA: Air Pollution Control Association, 1982. 119 pp. No ISBN.

This summary of the seminar, held in Washington, D.C., March 17–18, 1982, presents a review of air quality and trends in the United States. It focuses on the regulatory reform and international issues of the Clean Air Act amendments as well as on new directions for the EPA.

Bellini, James. **High Tech Holocaust.** Devon, Eng.: David and Charles, 1986. 255 pp. ISBN 0-7153-8812-6.
Evaluates health and environment hazards caused by industrial pollution.

Bhumralkar, C. M., and Jill Williams. **Atmospheric Effects of Waste Heat Discharges.** Energy, Power, and Environment no. 13. New York: M. Dekker, 1982. 157 pp. ISBN 0-8247-1653-1.
Although the demand for energy continues to increase, it is clear that there are constraints because of limited energy resources and the environmental impact of energy conversion. Some important effects of the production of electrical energy are the environmental changes caused by the production of waste energy—energy that generally appears as heat which is dissipated into the earth/atmosphere system. This book presents a comprehensive description of all aspects of the atmospheric effects of waste heat discharges from power plants as well as from major points of energy conversion such as urban and industrial areas.

Bretschmeider, Boris, and Jiri Kurfurst. **Air Pollution Control Technology.** Fundamental Aspects of Pollution Control and Environmental Science no. 8. New York: Elsevier, 1987. 296 pp. ISBN 0-444-98985-4.
This book provides an objective view of the importance of sources of atmospheric pollution and develops realistic possibilities for limitation of emissions. One major section stresses that the importance of air pollutants does not always correspond to magnitude. New chemicals that are particularly dangerous are continuously being developed, and when transported into the air, even in small amounts, they have concentrated temporal and spatial effects. The volume concludes with an economic analysis of controlling pollution, including a recognition of the fact that evaluation should include not only technical agents but also the cost of providing a sound environment.

Brimblecombe, Peter. **Air Composition and Chemistry.** Cambridge Environmental Chemistry Series. New York: Cambridge University Press, 1986. 224 pp. ISBN 0-521-25518-X.
A technical volume on the composition of the atmosphere and the chemical reactions that occur. Pollution problems are stressed.

Cannon, James S. **Controlling Acid Rain: A New View of Responsibility.** New York: INFORM, 1987. 53 pp. No ISBN.

This short volume discusses assigning responsibility for acid rain to the users or beneficiaries of midwestern coal-fired power.

Chigier, N. A. **Energy, Combustion, and Environment.** McGraw-Hill Series in Energy, Combustion, and Environment. New York: McGraw-Hill, 1981. 496 pp. ISBN 0-07-010766-1.
This book relates combustion science and fuel technology to energy production and utilization. Since all power systems have an impact on the atmosphere are examined, and practical means of control are recommended. The book is directed to scientists and engineers but will also be of interest to people involved in legislation and control of air pollution. A global viewpoint is maintained and a wide range of topics is discussed.

Conference of European Statisticians. **Statistics of Air Quality: Some Methods.** Statistics Standards and Studies no. 36. New York: UN, 1984. 49 pp. No ISBN.
This is a handbook of statistical methods for measuring atmospheric pollution.

Convention on Long-range Transboundary Air Pollution. **Air Pollution Across Boundaries.** Air Pollution Studies no. 2. New York: UN, 1985. 156 pp. ISBN 92-1-116328-5.
This volume presents a number of studies. The first considers the impact of air pollution on historical monuments. The second constitutes the first attempt by governments to estimate the extent and intensity of the present forest damage in an Economic Commission for Europe region-wide context. A third study presents evaluations of the efficiency and costs of technologies for controlling emissions of sulfur and nitric oxides from stationary sources. Fourth, there are reports on progress in elaborating a framework for an integrated assessment of air pollution control.

Cooper, C. David, and F. C. Alley. **Air Pollution Control: A Design Approach.** Boston, MA: PWS Engineering, 1986. 630 pp. ISBN 0-534-05910-4.
This book has two main objectives. The first is to present detailed information about air pollution and its control; the second concerns the formal design of equipment to remove impurities from the atmosphere. The text begins with a general understanding of air pollution and the design approach, but it quickly moves into a detailed, specific presentation of each major air pollution control system in use. Theory and practice are well balanced to provide a firm understanding of the subject.

Cooper, C. David, and F. C. Alley, eds. **Acid Rain Control II: The Promise of New Technology.** Papers and Proceedings of a

Conference Sponsored by the Illinois Energy Resources Commission and the Coal Extraction and Utilization Research Center, Southern Illinois University at Carbondale, April 10, 1985. Carbondale, IL: Southern Illinois University Press, 1986. 187 pp. ISBN 0-8093-1292-1.

This volume is a report on the second conference on acid rain held at Southern Illinois University. The papers included in the proceedings detail some of the latest developments in the search for understanding the problems of acid rain. Topics considered include the evolving science of acid deposition, renovation of acid lakes and streams, the rate of advanced coal-cleaning technology in the reduction of sulfur dioxide emissions, current technology for SO_2 emission control, and nuclear energy.

Cowling, Ellis B. **Pollutants in the Air and Acids in the Rain: Influence on Our Natural Environment and a Challenge for Every Industrial Society.** Horace M. Albright Conservation Lectureship, 1985. Berkeley, CA: University of California, College of Natural Resources, Department of Forestry and Resource Management, 1985. 26 pp. No ISBN.

In this lecture on the general problems of air pollution, three themes are developed: the pollutants in the air and acids in the rain, the influence of air pollutants on our natural environment, and the challenge they present to every industrialized society.

Elsom, Derek. **Atmospheric Pollution: Causes, Effects, and Control Policies.** New York: Basil Blackwell, 1987. 319 pp. ISBN 0-631-15674-7.

This study of atmospheric pollution examines its nature, sources, and effects on vegetation, crops, wildlife, man-made materials, buildings, and climate. Next is a discussion of possible strategies to combat the destruction of air pollution and a comparison of the control policies of the United States, the United Kingdom, the European Community, the Soviet Union, and China. The final section evaluates progress toward international collaboration on air pollution.

Georgii, H. W., and J. Pankrath, eds. **Deposition of Atmospheric Pollutants.** Proceedings of a Colloquium Held at Oberursel/Taunus, West Germany, November 9–11, 1981. Dordrecht, Neth.: D. Reidel Publishing Company, 1982. 217 pp. ISBN 90-277-1438-X.

The problem of acid precipitation has been recognized with growing concern in many industrialized countries, and research activities have been supported in many fields. In order to discuss the results of experimentation and theoretical work in the field of deposition and to gain a better understanding of the work done in different countries a colloquium was held in Oberursel/Taunus, West Germany, in November 1981. The proceedings are divided into three parts—dry deposition, wet deposition, and deposition on plants and vegetation—and present a good survey of present-day knowledge.

Gibson, Mary, ed. **To Breathe Freely: Risk, Consent, and Air.** Maryland Studies in Public Philosophy. Totowa, NJ: Rowman and Allanheld, 1985. 294 pp. ISBN 0-8476-7416-8.

Discusses risks that are _____

_____

Gilpin, Alan. **Air Pollution.** St. Lucia, Australia: University of Queensland Press, 1971. 67 pp. ISBN 0-7022-0717-9.

This volume presents a brief survey of the causes and effects of air pollution, major air pollution incidents, clean air legislation, air pollution control techniques, the economics of air pollution and control, and the location of industry.

Godish, Thad. **Air Quality.** Chelsea, MI: Lewis Publishers, 1985. 372 pp. ISBN 0-87371-019-3.

This is a comprehensive volume on air pollution as an interdisciplinary field, including chemistry, physics, meteorology, climatology, biology, political science, and economics, and provides a balanced study of this major environmental problem. Topics considered include atmospheric pollutants and effects, health effects, welfare effects, atmospheric surveillance, regulatory approaches, motor-vehicle emissions control, indoor air pollution, and noise pollution. The material is presented accurately, concisely, and lucidly with emphasis on the important aspects of the problem.

Graves, Philip E., and Ronald J. Krumm. **Health and Air Quality: Evaluating the Effects of Policy.** Studies in Economic Policy, AEI Studies 322. Washington, DC: American Enterprise Institute for Public Policy Research, 1981. 156 pp. ISBN 0-8447-3442-X.

Environmental policy in the United States has developed into a large body of stringent and complicated regulations. This study presents an overview of the issues associated with government regulation of air pollution and relates these issues to estimates of the cost and benefits that arise from regulations. The first part examines the economic content of pollution policies and their economic implications; the second focuses on the implications of nonlinear effects of single pollutants and the interactions among pollutants on a particular class of human health damages.

Greenberg, Michael R., ed. **Public Health and the Environment: The United States Experience.** New York: Guilford Press, 1987. 395 pp. ISBN 0-89862-778-8.

This volume presents a review of the major environmental problems today, drawing from both environmental health science and environmental policy. The authors, from a variety of fields, have experience in research, policy analysis, and the administration of government and private health and environmental agencies. This volume provides a background for evaluating the critical issues facing concerned citizens and policymakers.

Hay, Alastair. **The Chemical of Scythe, Lessons of 2,4,5-T, and Dioxin.** Disaster Research in Practice. New York: Plenum Press, 1982. 264 pp. ISBN 0-306-40973-9.

The principal goal of the Disaster Research in Practice series is to provide scientific and readable accounts about the most urgent areas of disaster research. This volume, dealing with the problems of chemical hazards, has chapters on dioxins, hexachlorophene, trichlorophenol, and chloraine. Case studies are presented for the dioxin explosion in Seveso, Italy, Vietnam and 2,4,5-T, and Love Canal. The final chapter considers how modern society can survive in a world of chemical hazards.

Henderson-Sellers, Brian. **Pollution of Our Atmosphere.** Bristol, Eng.: A. Hilger, 1984. 276 pp. ISBN 0-85274-754-3.

Pollutant emissions into the atmosphere change temporally and spatially as a result of technology, fuel usage, social attitudes, economic pressure, and many other factors. The existence of an air pollution problem results largely from the public's perception of the environment and the people's concern that existing technology could be used to improve the air quality of a locality. Air pollution, however, knows no political or geographical boundaries. The text covers all aspects of the subject: physics, observation techniques, health effects, fuel and combustion, prevention, control, and legislation.

Hesketh, Howard E. **Understanding and Controlling Air Pollution.** Ann Arbor, MI: Ann Arbor Science Publishers, 1972. 411 pp. ISBN 0-250-40007-3.

The concepts of ecology are fairly new to Americans. Only a few years ago it was thought necessary to exploit new frontier areas, but this exploitation must be stopped to regain a proper ecological balance. The first part of this volume considers air pollution and society, sources and emissions, pollution transport, air pollution chemistry and effects, and automotive pollution. The second part is devoted to controlling emissions and considers such engineering controls as classification of pollutants, combustion and related pollutants, particulate collection theory, control equipment, and the cost of air pollution control.

Horowitz, Joel L. **Air Quality Analysis for Urban Transportation Planning.** MIT Press Series in Transportation Studies no. 7. Cambridge, MA: MIT Press, 1982. 387 pp. ISBN 0-262-08116-4.

It has long been recognized that motor vehicles are a major contributor to urban air pollution. This volume deals exclusively with topics that are useful to persons involved in urban transportation planning and decision making, such as [illegible] and emissions control technology, and atmospheric dispersion. The orientation of the book is mainly scientific, technical, and analytical.

Industrial Guide for Air Pollution Control. Prepared for the Environmental Research Center by PEDCO Environmental, Inc. Technology Transfer/Handbook no. 4. Washington, DC: U.S. Environmental Protection Agency, Technology Transfer, 1978. 1 Vol. EPA-625/6-78-004. No ISBN.

This manual is intended for plant managers, engineers, and other personnel responsible for plant compliance with air pollution control regulations. Guidelines are established for plant emissions, emission regulations, and measurement and ambient air monitoring.

Key, David, and Harold K. Jacobson, eds. **Environmental Protection: The International Dimension.** Totowa, NJ: Rowman and Littlefield, 1983. 352 pp. ISBN 0-86598-034-9.

This comprehensive study explains the legal, social, and political ramifications of the dangers to human health posed by environmental damage and the potential economic consequences of resource destruction.

Lee, Robert E., Jr., ed. **Air Pollution from Pesticides and Agricultural Processes.** Cleveland, OH: CRC Press, 1976. 264 pp. ISBN 0-8493-5157-X.

It is now recognized that air pollution comes from many sources. This volume, prepared by a number of scientists, attempts to compile the available scientific information for understanding the complex subject of air pollution from pesticides and agricultural processes. The volume is divided into four parts. The first considers the chemical and physical aspects of airborne pesticides. The second is concerned with the toxicity of airborne pesticides. The third assesses the available legislative controls, and the fourth presents an overview of nonpesticidal air pollution from agricultural and food processing activities.

Legge, Allan H., and Sager V. Krupa, eds. **Air Pollutants and Their Effects on the Terrestrial Ecosystem.** Advances in

Environmental Science and Technology vol. 18. A Wiley-Interscience Publication. New York: Wiley, 1986. 662 pp. ISBN 0-471-08312-7.

The Advances in Environmental Science and Technology series is expected to stimulate interdisciplinary cooperation and understanding among environmental scientists. This volume, with 42 scientists participating, covers a wide range of subjects from government and industry regulatory perspectives, atmospheric chemistry, meteorological processes and pollutant deposition, pollutant measurement technology, vegetation and soil effects of air pollutants, and data acquisition strategies to total ecosystem research.

Marchello, J. M. **Control of Air Pollution Sources.** Chemical Processing and Engineering vol. 7. New York: Marcel Dekker, 1976. 638 pp. ISBN 0-8247-6182-0.

The control of air pollution sources has become a matter of vital importance to society, and this book is for scientists and engineers who are working in the field of air pollution control. The text covers air quality management, pollutant dynamics, pollutants in the atmosphere, particulate control equipment, gaseous pollutant control, control systems for energy conversion and manufacturing, and the cost of air pollution control.

Meetham, A. R. **Atmospheric Pollution: Its History, Origins, and Prevention.** 4th rev. ed. Pergamon International Library of Science, Technology, Engineering, and Social Studies. New York: Pergamon Press, 1981. 288 pp. ISBN 0-08-024003-8.

This basic book provides a comprehensive picture of air pollution. There is a logical progression of ideas from the origin of fossil fuels and their use in industrial boilers and furnaces and domestic heat services to atmospheric pollution, measurement of pollution, distribution of pollution, changes in pollution, effects of pollution, the prevention of atmospheric pollution, and legislation on air pollution.

National Research Council, Subcommittee on Airborne Particles. **Airborne Particles.** Subcommittee on Airborne Particles, Committee on Medical and Biologic Effects of Environmental Pollutants, Division of Medical Sciences, Assembly of Life Sciences, National Research Council, Medical and Biologic Effects of Environmental Pollutants. Baltimore, MD: University Park Press, 1979. 343 pp. ISBN 0-8391-0129-5.

This report considers those particles that are most often associated with general types of air pollution, and the emphasis is on particles that result from human activities. The types of particles and their distribution are considered, and the origin, behavior, and fate of such particles, their physical and chemical characteristics, and their interactions are discussed. Details are given on deposition, clearance, and retention of particles and their effect on people and animals.

Organization for Economic Cooperation and Development. **The Costs and Benefits of Sulphur Oxide Control: A Methodological Study.** Paris, Fr.: 1981. 164 pp. ISBN 92-64-12151-X.

A response to the work of the OECD Environment Committee on an overall air management policy, this study attempts to develop a methodology for a cost/benefit analysis of sulfur oxide control. It is divided into three main sections: an introduction, consisting of the background and philosophy of the study; a section describing how the impact on ambient air quality is estimated and how costs of control are calculated; and a final section covering an assessment of the benefits of sulfur oxide control for materials, crops, health, and aquatic ecosystems.

Painter, Dean E. **Air Pollution Technology.** Reston, VA: Reston Publishing Company, 1974. 283 pp. ISBN 0-87909-009-X.

This volume was written as an introductory college-level text for students who intend to become involved in air pollution control activities or for those who want to be informed about the fundamental problems of air pollution. The reader is introduced to the problems of air pollution by a discussion of the factors that contribute to air pollution, the types of pollutants, and the principal sources of air pollution. The effects of air pollutants on people, animals, vegetation, and nonliving materials are also covered. The discussion of atmospheric sampling and analysis is followed by chapters on control techniques and air resource management programs.

Perkins, Henry C. **Air Pollution.** New York: McGraw-Hill, 1974. 407 pp. ISBN 0-07-049302-2.

The basic objective of this volume is to provide information on certain specific pollutants, notably particulates, SO_2, and NO_X, as well as information on emissions. Other chapters deal with the effects of pollutants, both local and global; the history of federal legislative efforts; and enough background on combustion to understand many of the control techniques for combustion-generated pollutants. A substantial portion of the book deals with meteorology and the dispersion of pollutants from point sources.

Rowe, Robert D., and Lauraine G. Chestnut, eds. **Managing Air Quality and Scenic Resources at National Parks and Wilderness Areas.** Boulder, CO: Westview Press, 1983. 314 pp. ISBN 0-86531-941-3.

The enjoyment of a national park or recreational area may be affected by changes of visibility because of increased haze or by degradation of the landscape because of the destruction of natural features. This book discusses current research on methods to determine the value to visitors of preventing or mitigating the aesthetically undesirable impact on visual resources that may result from human activity. Topics discussed include human perceptions of air pollution, the role of air quality and visual resources, visual-resource management systems, and the application of social, psychological, and economic concepts of measuring the value of protecting visibility and visual resources.

Scorer, R. S. **Air Pollution.** New York: Pergamon Press, 1972. 151 pp. ISBN 08-013345-2.

This volume discusses the movement of air pollutants through the atmosphere. The basic mechanisms are considered, but the emphasis is upon seeing the mechanism in action, not theoretical formulas, so that by eye observation one can see what is happening with greater precision.

_____ . **Pollution in the Air: Problems, Policies, and Priorities.** London, Eng.: Routledge and Kegan Paul, 1973. 148 pp. ISBN 0-7100-7569-3.

This volume stresses that as industrial technologies have developed, so has air pollution. Difficult questions that must be considered include, Does increased technology always increase pollution? Is there a way out? Can we solve our problems without trying to stop industrial development? and Do we have to stop the growth of civilization to prevent that growth from becoming malignant? As a starting point the author recognizes that to develop technological control there must be public support and an understanding of the costs of such control.

Seinfeld, John H. **Air Pollution: Physical and Chemical Fundamentals.** New York: McGraw-Hill, 1975. 523 pp. ISBN 0-07-056042-0.

The basic aim of this book is to present in a rigorous, quantitative manner the fundamentals required for an analysis of the air pollution problem. The level is for advanced undergraduates and graduate students. Such topics are discussed as the elements of the air pollution problem, the origin and fate of air pollutants, air pollution meteorology, air pollution chemistry, micrometeorology, atmospheric diffusion, combustion processes, and air pollution control principles.

Sproull, Wayne T. **Air Pollution and Its Control.** 2d ed. New York: Exposition Press, 1972. 132 pp. ISBN 0-682-47490-8.

This is a general book on the technology of air pollution, particularly the invisible pollutants of oxides of sulfur and nitrogen, carbon monoxide, and hydrocarbons and the legal and economic aspects of the subject.

Stern, Arthur C., and others. **Fundamentals of Air Pollution.** 2d ed. New York: Academic Press, 1984. 530 pp. ISBN 0-12-666560-5.

This comprehensive volume presents the basic themes for the control of air pollution. The material is divided into four parts: the elements of air pollution, including the history of air pollution, air pollution systems, the scale of air pollution, and air quality; the effects of air pollution on inert materials, vegetation, animals, humans, and air quality criteria; the meteorology of air pollution; and the control of air pollution.

~~~~~~

Tolley, George S., and others. **Environmental Policy: Air Pollution.** Environmental Policy Series vol. 2. Cambridge, MA: Ballinger, 1982. 432 pp. ISBN 0-88410-626-8.

The basic purpose of this volume is to assist people who design and implement environmental policy to balance the social benefits of a policy action and the corresponding costs. The material is presented in three parts. Part one deals with the diverse benefits of cleaner air and considers the local, regional, and global concerns over health and property damage. Part two explores the various costs of alternative policy options such as added-on control devices, fuel switching, and output reductions. The policy implications of a decade of experience are brought together in part three, which provides a comprehensive review of the current state of knowledge regarding the formulation and analysis of air quality policies.

Treshow, Michael. **Whatever Happened to Fresh Air?** Salt Lake City, UT: University of Utah Press, 1971. 201 pp. ISBN 0-87480-062-5.

The objective of this book is to give the general reader information on air pollution. The work provides the framework to learn what air pollution is, what causes it, where it comes from, where it goes, what it does, what it costs, what officials are doing about it, and how it can be controlled. The reader can then evaluate the air pollution problems in his or her own community and make the correct decisions for its control.

United Nations, Economic Commission for Europe. **Air-borne Sulphur Pollution: Effects and Control.** Report prepared within the framework of the Convention on Long-Range Transboundary Air Pollution. Air Pollution Studies no. 1. New York: UN, 1984. 265 pp. No ISBN.

In this comprehensive study of airborne sulfur pollution, part one deals with the effects of sulfur compounds and related pollutants on natural life-supporting systems and on some materials. Growing evidence that insidious secondary effects are important elements in environmental backlashes, such as "first kills" and the "die back" of forests, makes the call for action urgent. The second part considers control technologies, and the final section considers a cost/benefit analysis of sulfur emission control.

Wark, Kenneth, and Cecil F. Warner. **Air Pollution: Its Origin and Control.** 2d ed. New York: Harper and Row, 1981. 526 pp. ISBN 0-7002-2534-X.

This volume is an introduction to air pollution problems for engineers and scientists. An understanding of the fundamentals of thermo-dynamics is assumed as well as knowledge of the basic concepts of chemical kinetics. The volume has chapters on the effects and sources of air pollutants, the dispersion of pollutants, particulates in the atmo-sphere, general control of gases, control of sulfur oxides and oxides of nitrogen, atmospheric photochemical reactions, mobile sources, and odor controls.

Particulate Pollution

Blong, R. J. **Volcanic Hazards: A Sourcebook on the Effects of Eruptions.** Orlando, FL: Academic Press, 1984. 440 pp. ISBN 0-12-107180-4.

This volume provides an excellent background to the topic of volcanic hazards.

Hesketh, Howard E. **Fine Particles in Gaseous Media.** 2d ed. Chelsea, MI: Lewis Publishers, 1986. 220 pp. ISBN 0-87371-030-4.

Particulate pollution is of major importance in the atmosphere, and this volume is a technical treatment of aerosols and the use of filters and filtration processes.

Pasquill, F., and F. B. Smith. **Atmospheric Diffusion: Study of the Dispersion of Windborne Material from Industrial and Other Sources.** Ellis Horwood Series in Environmental Science. New York: Halsted Press, 1983. 437 pp. ISBN 0-470-27404-2.

An effective dispersion of gaseous or finely divided material released into the atmosphere near the ground depends on natural mixing processes on a variety of scales. This book presents basic material on atmospheric diffusion, including discussions of turbulence and the atmospheric boundary layer, theoretical treatments of diffusion of materials, basic features of atmospheric diffusion, distribution of windborne materials, and estimation of local diffusion and dispersion over distances domi-nated by mesoscale and synoptic scale.

Schwartz, Stephen E., ed. **Trace Atmospheric Constituents: Properties, Transformations, and Fates.** Advances in Environmental Science and Technology vol. 12. A Wiley Interscience Publication. New York: Wiley, 1983. 547 pp. ISBN 0-471-87640-2.

This is a highly technical volume on atmospheric chemistry stressing meteorological inputs.

Seinfeld, John H. **Atmospheric Chemistry and Physics of Air Pollution.** A Wiley Interscience Publication. New York: Wiley, 1986. 738 pp. ISBN 0-471-82857-2.

The processes that impact species once they are released into the atmosphere involve a complete understanding of

the chemistry of air pollutants in the atmosphere; the formation, growth, and dynamics of aerosols; the meteorology of air pollutants; and the transport, diffusion, and removal of living matter in the atmosphere.

Sosa, Francisco J., Publio L. Fajardo, and Louis Theodore. Consulting editor, Anthony J. Buonicore. **Particulate Air Pollution; Problems and Solutions.** Boca Raton, FL: CRC Press, 1980. 101 pp. ISBN 0-8493-5541-9.

Presents the environmental aspects of particulate air pollution.

United Nations, Economic Commission for Europe. **Fine Particulate Pollution.** New York: Pergamon Press, 1979. 108 pp. ISBN 0-08-023399-6.

This volume summarizes the information obtained from a nine-part questionnaire submitted to scientists in 20 nations. The topics span all scientific and technical disciplines associated with atmospheric fine-particulate pollution—the definition of fine particulates, health effects, measurement, sources, transport and transformation, control technology, regulations, economics, and research. Improved measurement methods have greatly enhanced the sophistication of research on the physics and chemistry of atmospheric particulates and their biological effects.

Wise, William. **Air Pollution in Donora, Pa.: Epidemiology of the Unusual Smog Episode of October 1948.** United States Public Health Service Public Health Bulletin no. 306. Washington, DC: Federal Security Agency, Public Health Service, Bureau of State Services, Division of Industrial Hygiene, 1949. 173 pp. No ISBN.

Although the Donora, Pennsylvania, air pollution incident had tragic local significance, it also made the nation aware that atmospheric pollution could prove a serious health danger. This volume provides a biological and an atmospheric analysis of the problem and concludes with a discussion of the causes of the episode and recommendations for future control of industrial atmospheric pollutants.

Acid Precipitation

Acid Rain Information Clearinghouse (ARIC). **Acid Rain: Economic Assessment.** Edited by Paulette Mandelbaum. Environmental Science Research Series vol. 33. Rochester, NY: Plenum Press, 1985. 287 pp. ISBN 0-306-42102-X.

This volume presents the proceedings of a conference on acid rain, held in Washington, D.C., December 4–6, 1984, sponsored by the Acid Rain Information Clearinghouse. Explains uncertainties as well as areas of agreement that economists have established in using economics to better understand the nature of the acid rain problem.

_____. **Acid Rain: The Relationship Between Sources and Receptors.** Edited by James C. White. New York: Elsevier, 1988. 223 pp. ISBN 0-444-01277-X.

This book is designed for a nontechnical audience as well as for specialists in acid rain research. The chapters discuss the current legal status and the nature and scope of scientific understanding and research programs, identify areas of consensus and disagreement, and assess policy options in the light of current knowledge.

_____. **Liming Acidic Waters: Environmental and Policy Concerns.** Rochester, NY: Center for Environmental Information, 1985. 82 pp. No ISBN.

This book presents summaries of papers of a conference on liming acidic waters. The intent is neither to promote nor to discourage liming but to understand its ecological effects and to define its usefulness over the short term in treating damaged water systems.

Acid Rain 1986: A Handbook for States and Provinces— Research, Information, Policy. Proceedings of Wingspread Conference, held in Racine, Wisconsin, September 23–25, 1986, sponsored by the Johnson Foundation and the Acid Rain Foundation, Inc. St. Paul, MN: Acid Rain Foundation, 1986. 609 pp. ISBN 0-935577-07-6.

The objective of this conference was to update knowledge in three major areas of acid deposition: research, information, and policy on local, state or provincial, national, and international levels. Each state and province reported on each of those three areas. This volume surveys present-day scientific activities.

Acid Rain: Planning for the '80's. Contributing editors, Michael J. Zimmer and James A. Thompson, Jr. Rockville, MD: Government Institutes, 1983. 1 vol. (loose-leaf). ISBN 0-86587-118-3.

This volume presents a number of essays on a wide variety of acid rain topics: an analysis of the problem, planning horizons and the cost and financing implications, research on current technology, coal and acid rain, types of control of sulfur dioxides, the strategy of the Environmental Protection Agency, congressional and international responses, and present-day legal policy.

Boyle, Robert H., and R. Alexander Boyle. **Acid Rain.** New York:

...analysis of the scope of the acid rain problem and then discusses the widespread environmental effects of acidification with an emphasis on the destruction of water and forest resources. The difficulty of developing a potential position to control pollution is stressed, and the reluctance of industry to recognize the problem is also emphasized. The authors believe that the problems of acid precipitation can be solved with more effective political control.

Bubenick, David V. **Acid Rain Information Book.** 2d ed. Park Ridge, NJ: Noyes Publications, 1984. 397 pp. ISBN 0-8155-0967-7.

This book is the result of the combined efforts of many staff members of the GCA/Technology Division, and it discusses the major aspects of the acid rain problem in existence today. It points out areas of uncertainty and summarizes current projected research. The volume is organized in a logical progression from the sources of pollutants that lead to acid rain formation to the atmospheric transport and transformation of those pollutants and finally to the deposition of acid rain, the effects of that deposition, monitoring and modeling procedures, and possible mitigative measures and regulatory options.

Calvert, Jack, and others. **Acid Deposition, Atmospheric Processes in Eastern North America: A Review of Current Scientific Understanding.** Washington, DC: National Academy Press, 1983. 375 pp. ISBN 0-309-03389-6.

This volume presents a review of advances in understanding the problems of acid deposition in eastern North America. The meteorological aspects are stressed.

Cannon, James S. **Controlling Acid Rain: A New View of Responsibility.** New York: INFORM, 1987. 53 pp. ISBN 0-918780-41-1.

This study assigns responsibility for acid rain and its methods of control to the midwestern coal-fired plants of the United States. Special emphasis is given to the benefits of controlling acid rain.

Canter, Larry W. **Acid Rain and Dry Deposition.** Chelsea, MI: Lewis Publishers, 1986. 370 pp. ISBN 0-87371-016-9.

This volume discusses the physical and chemical aspects of wet and dry deposition.

Carr, Donald E. **The Sky Is Still Falling.** New York: W. W. Norton and Company, 1982. 278 pp. ISBN 0-393-01508-4.

A review of acid rain environmental problems as they have evolved in recent times.

Carroll, John E. **Acid Rain: An Issue in Canadian-American Relations.** Washington, DC: Canadian-American Committee, 1982. 80 pp. ISBN 0-89068-064-7.

A short report on the issue of acid rain in the relations between Canada and the United States.

Ciolkosz, Edward J., and Elissa R. Levine. **Evaluation of Acid Rain Sensitivity of Pennsylvania Soils.** Research Project Technical Completion Report A-058-PA. University Park, PA: Institute for Research on Land and Water Resources, Pennsylvania State University, 1983. 106 pp. No ISBN.

This study presents the results of a computer simulation model to determine the impact of various inputs of acid deposition on Pennsylvania soils. The model simulated the changes that would occur in the solid phase of soils in humid, temperate climates undergoing acidification and cation leaching. The state was divided into ten regions, and soils were classified into various sensitivity classes. The model showed that the soils of Pennsylvania were either very sensitive or nonsensitive and that the nonsensitive soils contained sufficient buffer capacity to withstand acid deposition.

Crocker, Thomas D., ed. **Economic Perspectives on Acid Deposition Control.** Acid Precipitation Series vol. 8. Boston, MA: Butterworth Publishers, 1984. 180 pp. ISBN 0-250-40565-2.

Consisting of papers given at the 1982 meeting of the American Chemical Society, the basic goal of this volume is to present viewpoints on the economics of controlling acid rain deposition. Topics treated include the effect of global optimization on locally optimal pollution control; the economic effects of acid rain; legal, ethical, economic, and political aspects of transfrontier pollution, acidification impact on fisheries, and transferable discharge permits; and profit-maximizing behavior.

Dudley, Nigel, Mark Barrett, and David Baldock. **The Acid Rain Controversy.** London, Eng.: Earth Resources Research, 1985. 177 pp. ISBN 0-946281-09-2.

This volume presents the long-held British viewpoint that casts doubt on the alleged extent of acidification and the degree to which it is attributable to air pollution. The bulk of the scientific work on acidification in Britain has been undertaken by the Central Electricity Generating Board, by a large margin the island's largest emitter of sulfur dioxide. Britain has argued against international control and believes that more research is required before action can occur. Most European countries are opposed to the British position, which reflects national self-interest as converting to nonpolluting fuels would be costly.

Sourcebook. New York: McGraw-Hill, 1984, 290 pp. ISBN 0-07-606540-5.

This volume publishes papers from the First International Conference on Acid Rain held in Washington, D.C., March 27–28, 1984. The volume is divided into seven parts and 20 chapters plus an overview designed to give readers a brief review of the problem of acid rain. The material covers the problem and legislative solutions; international mitigation programs; the planning of U.S. programs; emission reduction before, during, and after combustion; and engineering solutions under development.

Elsworth, Steve. **Acid Rain.** London, Eng.: Pluto Press, 1984. 154 pp. ISBN 0-86104-791-5.

A general study of the growth of acid rain deposition with a discussion of the environmental impact on Europe.

Fraser, G. Alex, and others. **The Potential Impact of the Long Range Transport of Air Pollutants on Canadian Forests.** Information Report no. E-X-36. Edmonton, Alta.: Canadian Forestry Service, University of Alberta, 1985. 43 pp. No ISBN.

This study sheds considerable light on the effect on forests of the long-range transport of air pollutants. Using an iterative series of four questionnaires, expert opinion was solicited on the nature and extent of forest productivity change and the likelihood of different rates of forest productivity effects under several different pollution scenarios. The results provide a realistic picture of the risks to Canadian forests from present and potential future levels of pollution.

Gilleland, Diane Suitt, and James H. Swisher, eds. **Acid Rain Control: The Costs of Compliance.** Papers and proceedings of a conference sponsored by the Illinois Energy Resources Commission and the Coal Extraction and Utilization Research Center, Southern Illinois University at Carbondale, Carbondale, March 28, 1984. Carbondale, IL: Southern Illinois University Press, 1985. 177 pp. ISBN 0-8093-1205-0.

The proceedings of this conference, held to make the state of Illinois aware of coal research needs and to give Southern Illinois University a leadership role in that research, considered a wide variety of topics, including acid deposition, problems of modeling the impacts of acid rain legislation, employment effects of proposed acid rain legislation, controlling acid rain, and compliance costs.

Howard, Ross, and Michael Perley. **Acid Rain: The North American Forecast.** Toronto, Ont.: Anansi, 1980. 206 pp. ISBN 0-88784-082-5.

This volume presents a number of graphic descriptions of the effects of acid rain on the environment. One of the most interesting is the analysis of the dying lakes; other aspects include acid in the air, the ecosystem, and the community. This section is followed by chapters on the economics of acid rain; the political aspects, centering on the problems of legislation and controls between the United States and Canada; and forecast of environmental degradation.

Kahan, Archie M. **Acid Rain: Reign of Controversy.** Golden, CO: Fulcrum, 1986. 238 pp. ISBN 1-55591-003-3.

This volume analyzes the controversy concerning the origin of acid rain.

Katzenstein, Alan W. **An Updated Perspective on Acid Rain.** Prepared for the EEI Acid Rain Public Response Task Force. Washington DC: Edison Electric Institute, 1981. 44 pp. No ISBN.

This book is a primer on acid rain. It provides basic material such as defining acid rain, explaining pH, the nature of clean air, measurement of pH, trends of acid rain development, long-range transport, effects of acidity, corrective strategies, and what is being done about the problem.

Luoma, Jon R. **Troubled Skies, Troubled Waters: The Story of Acid Rain.** New York: Viking Press, 1984. 178 pp. ISBN 0-1400-8094-5.

This volume is intended to inform the general public of the danger of acid precipitation. The material is presented in a conversational manner in nontechnical language, and many personal accounts provide a realism to the problems of acid rain. Discusses not only the problems but also the difficulty of providing satisfactory solutions.

Lynch, James A., Edward S. Corbett, and Kevin M. Kostelnik. **Atmospheric Deposition: Spatial and Temporal Variation in Pennsylvania—1986.** Prepared for the Department of Environmental Resources. University Park, PA: Environmental Resources Research Institute, Pennsylvania State University, 1987. 119 pp. No ISBN.

This report presents a summary of precipitation chemistry and wet deposition data collected throughout Pennsylvania during 1986. Also included are data on the concentrations of ions found in dry deposition. The report concludes that precipitation in Pennsylvania can be characterized as a dilute, aqueous solution of sulfuric and nitric acid. There were also measurable concentrations of chloride, ammonium, calcium, potassium, sodium, and magnesium. Dry deposition also occurred in the state in 1986.

Pollution. Edited by Jon Tinker. Washington, DC: International Institute for Environment and Development, 1985. 191 pp. ISBN 0-905347-61-7.

This volume first presents background chapters on the formation of acids, acid drainage, and reducing and controlling acid pollution. It then stresses that acid pollution is an international issue, and there are chapters devoted to Scandinavia, West Germany, Britain, other European countries, Eastern Europe and the Soviet Union, North America, Japan, the Third World, and the Arctic. The book is a good world survey.

National Research Council, Committee on Monitoring and Assessment of Trends in Acid Deposition, Environmental Studies Board, Commission on Physical Sciences, Mathematics, and Resources. **Acid Deposition Long-Term Trends.** Washington, DC: National Academy Press, 1986. 506 pp. ISBN 0-309-03647-X.

Because the deposition of chemical pollutants from the atmosphere constitutes one of the major environmental issues of our time, the National Research Council has prepared a series of studies designed to summarize what is known about this complex phenomenon. A central objective has been to delineate more clearly the role of industrialized society in contributing to the chemical composition of the atmosphere. This study addresses the dimension of the relationships among emissions, deposition, and environmental effects.

Ostmann, Robert, Jr. **Acid Rain: A Plague Upon the Waters.** Minneapolis, MN: Dillon Press, 1982. 208 pp. ISBN 0-87518-224-0.

This volume presents a comprehensive view of the acid rain problem. After reviewing how acid rain is a threat to the world, the author presents chapters on the loss of heritage, the Scandinavian connection, and the pioneers who studied the effects of acid rain. The final chapters are devoted to difficulties of attacking the problem of acid rain. The volume is written in simple, nontechnical language.

Overrein, Lars N., Hans Martin Seip, and Arne Tollan. **Acid Precipitation—Effects on Forest and Fish.** Final Report of the

SNSF-Project 1972-1980. 2d ed. Fagrapport (Forsknings-prosjekt Sur Nedbors Virkning Pa Skog Og Fisk), FR 19/80. Oslo, Nor.: SNSF Project, 1981. 175 pp. ISBN 82-90376-16-2.

The basic purpose of this volume is to establish as precisely as possible the effects of acid precipitation on forests and freshwater fish and to investigate the effects of air pollutants on soil, vegetation, and water. An interdisciplinary study brought together the capabilities of different Norwegian research institutes, and this final report summarizes the result of some 300 articles and data presented in various international journals between 1972 and 1980.

Pawlick, Thomas. **A Killing Rain: The Global Threat of Acid Precipitation.** San Francisco, CA: Sierra Club Books, 1984. 206 pp. ISBN 0-87156-823-3.

This readable account of the effects of acid rain on the environment presents the problems through stories of how acid rain has affected individuals. In this manner an effective case is made showing how destructive acid rain can be to society as a whole. The message is that the present systems of control provide only temporary solutions; the development of a basic source of nonpolluting energy is the only permanent solution.

Pearce, Fred. **Acid Rain.** London, Eng.: Penguin Books, 1987. 162 pp. ISBN 0-14-052380-4.

This basic text explains what acid rain is and the problems that are being created in the environment. The material stresses the problem in Europe.

Postel, Sandra. **Air Pollution, Acid Rain, and the Future of Forests.** Worldwatch Paper 58. Washington, DC: Worldwatch Institute, 1984. 54 pp. ISBN 0-916468-57-7.

This volume presents a comprehensive picture of the effects of acid rain on the forests of the world. Topics discussed include signs of forest destruction, tracing the pathways of pollution, the cost of acid rain, cutting emissions, and the survival of the forests.

Proceedings of Mid-South Symposium on Acid Deposition, Little Rock, Arkansas, April 20-21, 1986. Edited by B. G. Blackmon and R. S. Beasley. Monticello, AR: University of Arkansas, 1986. 62 pp. No ISBN.

This volume consists of seven articles on different aspects of the acid rain problem throughout the world, including reports on atmospheric deposition in the United States, atmospheric deposition and cation leaching, the relationship between forest decline and soil and water acidification in Scandinavia and northern Germany, the acidification of lakes in the eastern part of the United States, and atmospheric deposition and research needs and priorities in the mid-South.

Raufer, Roger K., and Stephen L. Feldman. **Acid Rain and Emissions Trading: Implementing a Market Approach to Pollution Control.** Totowa, NJ: Rowman and Littlefield, 1987. 200 pp. ISBN 0-8476-7555-6.

A key policy proposal in the ~~~~~~~~~~~~~~~~~~~

rights. This book is a comprehensive study of the theoretical and practical difficulties involved in the creation of a national emissions trading program, a survey of the positions of the utilities and public regulatory commissions toward such a program, and an estimation of the potential environmental and economic impacts on emissions trading.

Ribblett, Gary C., David R. DeWalle, and J. David Helvey. **Chemistry of Leachate from Six Different Appalachian Forest Floor Types Subjected to Simulated Acid Rain.** Prepared for USDA, Forest Service, Northeastern Forest Experiment Station, Timber and Watershed Laboratory, Parsons, West Virginia. University Park, PA: Pennsylvania State University, Institute for Research on Land and Water Resources, 1982. 40 pp. No ISBN.

A technical report on the chemistry of leachates on forest floors in the Appalachian region that were subjected to simulated acid rain.

Roth, Philip, and others. **The American West's Acid Rain Test.** Washington, DC: World Resources Institute, 1985. 50 pp. ISBN 0-915825-07-4.

This report considers acid deposition in the western part of the United States. Although the eastern part has been often studied with regard to the causes and effects of acid deposition, the West has had little attention. This study recognizes that the chemical deposition in the West is different, has a different transport pattern, and has different ecological impacts requiring different control strategies. This study evaluates the problems and recommends preventive measures.

Schwieger, Robert G., and Thomas C. Elliott, eds. **Acid Rain Engineering Solutions, Regulatory Aspects.** New York: Hemisphere Publishing and McGraw-Hill, 1985. 1 vol. (various paging). No ISBN.

This volume discusses the technological aspects of the control of acid rain, stressing the environmental aspects of the sulfide and nitric oxides. As a conclusion, environmental policy is evaluated.

Smith, Oliver F., Jr. **An Assessment of Acid Rain.** Philadelphia, PA: Pennsylvania Environmental Research Foundation, 1982. 63 pp. No ISBN.

A general discussion of the recent development of acid rain.

Smith, William H. **Air Pollution and Forests: Interactions Between Air Contaminants and Forest Ecosystems.** Springer Series on Environmental Management. New York: Springer-Verlag, 1981. 379 pp. ISBN 0-387-90501-4.
This volume provides a compendium of the most significant relationships between forests and air pollution under low, intermediate, and high acid rain conditions. As the dose rises, the injury to the forest increases until the exposure to high doses may induce acute morbidity or mortality of specific trees. Such topics are discussed as the role of forests in major clement cycles, forests as a sink for air contaminants, forest tree reproduction, forest nutrient cycling, forest metabolism, forest stress, and forest ecosystem destruction.

Van Lier, Irene H. **Acid Rain and International Law.** Toronto, Ont.: Publishing Division, Bunsel Environmental Consultants, 1981. 278 pp. Based on LL.M. thesis, Dalhousie University. No ISBN.

Because acid rain deposition is international in scope, international laws must be developed to cope with this environmental problem. This volume stresses the problems of the Canadian-U.S. legal controversy.

Weller, Phil. **Acid Rain: The Silent Crisis.** Kitchener, Ont.: Between the Lines and the Waterloo Public Interest Research Group, 1980. 94 pp. ISBN 0-919946-16-X.

This volume deals not only with the scientific and environmental problems of acid rain but also with the human consequences. In order to understand the problems of acid rain, the political, social, and economic factors must be explored, and much of this book is devoted to examining the decisions by industry and government that led to the persistence of acid rain. Only after acquiring this type of information can the necessary programs be developed to alleviate acid rain damage.

Yanarella, Ernest J., and Randal H. Ihara, eds. **The Acid Rain Debate: Scientific, Economic, and Political Dimensions.** Boulder, CO: Westview Press, 1985. 342 pp. ISBN 0-8133-7065-5.

This volume presents a variety of viewpoints that have developed as a result of the controversy as to the origin of acid rain. U.S. policy issues are stressed and compared with Canadian and West European viewpoints.

Carbon Dioxide

Breuer, Georg. **Air in Danger: Ecological Perspectives of the Atmosphere.** Translated by Peter Fabian. Rev. English ed. New York: Cambridge University Press, 1980. 189 pp. ISBN 0-521-22417-9.

Largely limited to the importance of CO_2 in the atmosphere, this volume describes how the atmosphere, and the processes regulating its composition have developed the ... ... processes. It analyzes how and to what extent this natural equilibrium might be disturbed by human interference and what might be the consequences for life on earth.

Mintzer, Irving. **A Matter of Degrees: The Potential for Limiting the Greenhouse Effect.** Washington, DC: World Resources Institute, 1987. 113 pp. ISBN 0-915825-25-2.

In late 1985 scientists at an international conference warned that increases of carbon dioxide and other greenhouse gases in the atmosphere could raise the temperature of the earth 0.5°–1.5° C over preindustrial temperatures. To test this hypothesis the author integrated existing simulation models into one structure that predicts the emission of the six gases that contribute most to global warming. The author concludes that strong measures can significantly diminish the buildup of greenhouse gases if implemented soon.

Ozone

Miller, Alan, and Irving Mintzer. **The Sky Is the Limit: Strategies for Protecting the Ozone Layer.** Washington, DC: World Resources Institute, 1987. 43 pp. ISBN 0-915825-17-1.

This volume reviews the current understanding of the risk of ozone modification. It describes the techniques for reducing and eliminating emission of CFCs and addresses several key issues before the United States and other nations. Emphasis is placed on the alternatives that are available and on the fact that the cost of protecting the ozone is modest. The key issue is marshaling the political will before the alternatives disappear and the costs rise dramatically.

Radioactive Contamination

Marples, David R. **Chernobyl and Nuclear Power in the U.S.S.R.** Basingstoke, Eng.: Macmillan Press in association with Canadian Institute of Ukrainian Studies, University of Alberta, 1987. 228 pp. ISBN 0-312-00457-5.

This volume reviews the place of nuclear power in the Soviet Union and the importance of the Chernobyl accident in that nation.

Photochemical Air Pollution

National Research Council, Committee on Medical and Biologic Effects of Environmental Pollutants. **Ozone and Other Photochemical Oxidants.** Medical and Biologic Effects of Environmental Pollutants series. Washington, DC: National Academy of Sciences, 1977. 719 pp. ISBN 0-309-02531-1.

A scientific discussion of the biological effects of environmental pollutants.

Photochemical Smog and Ozone Reactions. Advances in Chemistry series no. 113. Washington, DC: American Chemical Society, 1972. 285 pp. No ISBN.

A symposium sponsored by the Division of Physical Chemistry, Industrial and Engineering Chemistry, and Water, Air and Waste Chemistry of the American Chemical Society at the 161st ACS meeting. Technical treatment of the origin of smog.

Photochemical Smog: Contribution of Volatile Organic Compounds. Washington, DC: Organization for Economic Cooperation and Development, 1982. 98 pp. ISBN 92-64-12297-4.

A chemical analysis of the organic components of smog.

Symposium on Chemical Reactions in Urban Atmospheres, 1969, Warren, Michigan. **Chemical Reactions in Urban Atmospheres, Proceedings.** Edited by Charles S. Tuesday. New York: American Elsevier, 1971. 287 pp. No ISBN.

An analysis of chemical reactions in urban atmospheres.

Indoor Pollution

Indoor Air Quality Environmental Information Handbook: Radon. Prepared by Mueller Associates, Syscon Corporation, and Brookhaven National Laboratory for U.S. Department of Energy, Assistant Secretary for Environment, Safety and Health, Office of Environmental Analysis. Washington, DC: Office of Environmental Analysis, January 1986. 1 vol. (various paging). No ISBN.

As homeowners have decreased ventilation rates inside their homes in an effort to conserve energy, the level of indoor air pollutants has been affected. The indoor accumulation of radon, a naturally occurring, radioactive gas, is related to ventilation. The Office of Environmental Analysis has prepared this handbook in an effort to bring together available information on the impact of radon and its decay products on residential indoor air quality and on human health. The material includes a comprehensive review of source materials, sources of radon, transport mechanisms, indoor concentrations, health effects and standards, and control technologies

International Speciality Conference Sponsored by APCA's TT-7 Indoor Air Quality Committee and hosted by the Delaware Valley Chapter of APCA's Mid-Atlantic States Section, Philadelphia, Pennsylvania, February 1986. Pittsburgh, PA: Air Pollution Control Association, 1986. 252 pp. No ISBN.

The purpose of the APCA conference was to review past experiences and to characterize the current state of knowledge on measurement and control of indoor radon in the United States, Canada, and other countries. This volume consists of the papers given at that conference.

National Research Council, Assembly of Life Sciences. **Indoor Pollutants.** Washington, DC: National Academy Press, 1981. 537 pp. ISBN 0-309-03188-5.

Because human beings spend so much time indoors, the present quality of the indoor environment and how that quality may change are matters of concern, and this volume reviews the problems of human exposure to indoor pollutants. Although there is little epidemiologic evidence of the health effects of indoor pollutants, it has been shown that some pollutants exceed ambient air quality standards. This book discusses the character of indoor pollution, monitoring, and modeling; factors that affect exposure to indoor pollution; the health effects of indoor pollution; effects on human welfare; and the control of indoor pollution.

Recent Advances in the Health Sciences and Technology. Edited by Birgitta Berglund, Thomas Lindvall, and Jan Sundell. Third International Conference on Indoor Air Quality and Climate, Stockholm, 1984. Stockholm: Swedish Council for Building Research, 1984. 284 pp. No ISBN.

This volume presents a wide range of viewpoints on the problems and effects of indoor pollution on humans. The topics include the lung cancer risk from radon daughter exposure; air quality control strategies for health, comfort, and energy efficiency; the potential of lung cancer from passive smoking; evaluation of thermal discomfort; the legal effects of formaldehyde off-gassing; and airborne injections and modern building technology.

Laws and Regulations

Ackerman, Bruce A., and William T. Hassler. **Clean Coal/Dirty Air: Or, How the Clean Air Act Became a Multibillion-Dollar Bail-Out for High-Sulfur Coal Producers and What Should Be Done About It.** New Haven, CT: Yale University Press, 1981. 193 pp. ISBN 0-300-02628-5.

This volume begins with an analysis of the environmental legislation of the late 1960s and 1970s and then focuses upon the crucial substantive policy issues regarding the future of the coal-burning power plant. Emphasis is placed on the way the U.S. Congress attempted to control the environment and the EPA's response in creating a system of controls. The volume concludes by exploring the implications of the EPA's decision for the future of environmental law.

American Enterprise Institute for Public Policy Research. **The Clean Air Act: Proposals for Revisions: 97th Congress, 1st Session, 1981.** Washington, DC: 1981. 90 pp. ISBN 0-8447-0245-5.

The effectiveness of the Clean Air Act and the proposals for revision to make it more effective are discussed.

Friedman, James M., and Michael S. McMahon. **The Silent Alliance: Canadian Support for Acid Rain Controls in the United States and the Campaign for Additional Electricity Exports.** Chicago, IL: Regnery Gateway, 1984. 78 pp. ISBN 0-89526-603-2.

This volume presents the Canadian viewpoint for control of acid rain originating in the United States and the desire of Canada to export electricity based on that country's waterpower resources.

Gould, Roy. **Going Sour: Science and Politics of Acid Rain.** Cambridge, MA: Birkhäuser Boston, 1985. 155 pp. ISBN 0-8176-3251-4.

This book presents a scientific review and policy analysis of acid rain. It reviews such questions as the roots of the problem and how they developed, how acid rain affects the environment, and what are the social, economic, and political consequences of solving the problems of acid rain. The volume briefly examines the role of the Environmental Protection Agency and the courts in administering the laws governing air pollution. The material is presented in a nontechnical manner.

Haskell, Elizabeth H. **The Politics of Clean Air: EPA Standards for Coal-Burning Power Plants.** Praeger Special Studies/Praeger

Scientific. New York: Praeger Publishers, 1982. 206 pp. ISBN
0-03-059701-3.

This book is a case study for the regulation of clean air. It documents the
U.S. Environmental Protection Agency's decision process in setting air
pollution control standards for future coal-fired power plants to be built
by electric utility companies, and it analyzes air pollution, coal utilization and the utility industry and its politics, and policies of federal
regulation. The book is organized chronologically f~~~ ~~~~~~~~

Lave, Lester B., and Gilbert S. Omenn. **Cleaning the Air:
Reforming the Clean Air Act: A Staff Paper.** Studies in the
Regulation of Economic Activity. Washington, DC: Brookings
Institution, 1981. 51 pp. ISBN 0-8157-5159-1.

The authors of this publication argue that most of the air pollution
abatement in the postwar period is attributable to economically induced
fuel switching rather than to enforcement of the Clean Air Act. That act
has not been effective because Congress mandated that it be implemented by administrative and legal procedures emphasizing due process
rather than efficacy and efficiency. The volume emphasizes the need to
place greater reliance on economic incentives. Congress should delegate
the details of regulation without relinquishing its oversight responsibility, specify the goals to be achieved, and ensure the scientific integrity of
the regulatory process. Finally, the authors argue that air pollution
monitoring and inspection systems are in shambles and need congressional attention.

Lundqvist, Lennart J. **The Hare and the Tortoise: Clean Air
Policies in the United States and Sweden.** Ann Arbor, MI:
University of Michigan Press, 1980. 248 pp. ISBN 0-472-09310-X.

This volume is a comprehensive study of clean air policy development
and change in the United States and Sweden. Emphasis is placed on
how and why such policies have developed and changed, and the basic
concern is with the relationship between policy content and the cultural,
economic, social, and political factors that are usually believed to influence policy content. Of the three parts, the first describes and compares
the major clean air policy choices made in the two countries in 1969–1970.
The second part analyzes the intentions, motivations, and calculations
of the policymakers at the time and looks at the arguments for or against
the chosen major controls, the allocation of policy authority, and the
mechanisms of public participation. The third part traces the changes in
clean air policies during the 1970s.

Melnick, R. Shep. **Regulation and the Courts: The Case of the
Clean Air Act.** Studies in the Regulation of Economic Activity.

Washington, DC: Brookings Institution, 1983. 404 pp. ISBN 0-8157-5662-3.

This volume examines how the federal courts have influenced policymaking in one highly significant and controversial area, the regulation of air pollution. The long-term effects of the variety of court decisions that helped to shape the environmental policies during the 1970s are evaluated, and the analysis of the bureaucratic and congressional politics of air pollution control provides insight into the process of environmental regulation as well as the effects of judicial activism. The author emphasizes that court decisions have consequences unforeseen by judges and legal commentators and also that the net effect of a large number of trial and appellate court decisions has been to widen the gap between the performance of the programs administered by the Environmental Protection Agency.

Munton, Don, and Susan Eros, eds. **Political and Legal Aspects of a Canada-United States Air Quality Accord.** Proceedings of a Workshop, Toronto, Canada, December 17, 1981. Toronto, Ont.: Canadian Institute of International Affairs, 1982. 90 pp. ISBN 0-919084-42-7.

This volume outlines the differences in Canadian and U.S. viewpoints concerning acid rain control and analyzes the political and legal aspects of acid rain.

United Nations Environmental Programme, Governing Council, 1983, Eleventh Session, Nairobi. **The State of the Environment 1983: Selected Topics.** Nairobi, Kenya: United Nations Environment Programme, 1983. 38 pp. ISBN 92-807-1076-1.

This small volume discusses three topics of international significance selected by the Governing Council of the United Nations Environmental Programme: hazardous waste, acid rain, and environmental aspects of energy farms. The essays provide much basic information on these environmental problems.

White, Lawrence J. **The Regulation of Air Pollutant Emissions from Motor Vehicles.** Washington, DC: American Enterprise Institute for Public Policy Research, 1982. 110 pp. ISBN 0-8447-3492-6.

The federal program to regulate air pollutant emissions from mobile sources has been an important part of the federal government's overall strategy for reducing air pollution in the United States. This monograph reviews the federal mobile source program and analyzes its current provisions. The argument is made that there has been a substantial reduction in emissions but that the costs imposed on vehicle users have been substantial. The focus is on the federal program, but some attention is given to California's more stringent program.

Journal Articles
and Government Documents

"Air Quality and Urban Development." *Environmental Comment (Urban Land Institute)* (April 1980): 3–16.

"An Appraisal of Louisiana's Air Quality Program: A Summary." *PAR (Public Affairs Research) Analysis* (November 1982): 1–11.

Barrie, L. A. "Air Pollution: An Overview of Current Knowledge." *Atmospheric Environment* 20: 4 (1986): 643–663.

Bowling, S. A. "Climatology of High-Air Pollution as Illustrated by Fairbanks and Anchorage, Alaska." *Journal of Climate and Applied Meteorology* 25 (January 1986): 22–34.

Brown, L. M., and others. "Lung Cancer in Relation to Environmental Pollutants Emitted from Industrial Sources." *Environmental Research* 34 (August 1984): 250–261.

Busch, R. H., and others. "Effects of Ammonium Nitrate Aerosol Exposure on Lung Structure of Normal and Elastase-Impaired Rats and Guinea Pigs." *Environmental Research* 39 (April 1986): 237–252.

Call, Gregory D. "Arsenic, ASARCO, and EPA: Cost-Benefit Analysis, Public Participation, and Polluter Games in the Regulation of Hazardous Air Pollutants." *Ecology Law Quarterly* 12:3 (1985): 567–617.

Chappie, Mike, and Lester Lave. "The Health Effects of Air Pollution: A Reanalysis." *Journal of Urban Economics* 12 (November 1982): 346–376.

Chung, Y. S. "Air Pollution Detection by Satellites: The Transport and Deposition of Air Pollutants over Oceans." *Atmospheric Environment* 20:4 (1986): 617–630.

Cross, Frank B. "Section 111(d) of the Clean Air Act: A New Approach to the Control of Airborne Carcinogens." *Boston College Environmental Affairs Law Review* 13:2 (1986): 215–240.

Drummond, J. G., and others. "Comparative Study of Various Methods Used for Determining Health Effects of Inhaled Sulfates." *Environmental Research* 41 (December 1986): 514–528.

Evans, Gary W., and Stephen V. Jacobs. "Air Pollution and Human Behavior." *Journal of Social Issues* 37:1 (1981): 95–125.

Halder, C. A., and others. "Gasoline Vapor Exposures." *American Industrial Hygiene Association Journal* 47 (March 1986): 164–175.

Kawamura, K., and I. R. Kaplan. "Biogenic and Anthropogenic Organic Compounds in Rain and Snow Samples Collected in Southern California." *Atmospheric Environment* 20:1 (1986): 115–124.

Kentucky, Legislative Research Commission. *The Measurement of Air Quality in the Commonwealth.* Charles Hardin and Peggy Hyland. Research Report no. 179. Frankfort, KY: 1981. 105 pp.

Logsdon, Jeanne M. "Organizational Responses to Environmental Issues: Oil Refining Companies and Air Pollution." In *Research in Corporate Social Performance and Policy,* edited by Lee E. Preston, pp. 47–71. Greenwich, CT: JAI Press, 1985.

McClellan, R. O. "Health Effects of Diesel Exhaust: A Case Study in Risk Assessment, 1985 Stokinger Lecture." *American Industrial Hygiene Association Journal* 47 (January 1986): 1–13.

Melnick, R. Shep. "Pollution Deadlines and the Coalition for Failure." *Public Interest (New York)* (Spring 1984): 123–134.

"Protecting Our Air." *EPA (Environmental Protection Agency) Journal* 10 (September 1984): 2–25.

Rotton, James, and James Frey. "Psychological Costs of Air Pollution: Atmospheric Conditions, Seasonal Trends, and Psychiatric Emergencies." *Population and Environment* 7 (Spring 1984): 3–16.

Sattar, S. A., and M. K. Ijaz. "Spread of Viral Infections by Aerosols." *Critical Reviews in Environmental Control* 17:2 (1987): 89–131.

Schlesinger, R. B. "Comparative Irritant Potency of Inhaled Sulfate Aerosols—Effects on Bronchial Mucociliary Clearance." *Environmental Research* 34 (August 1984): 268–279.

Seskin, Eugene P., and others. "An Empirical Analysis of Economic Strategies for Controlling Air Pollution." *Journal of Environmental Economics and Management* 10 (June 1983): 112–124.

Spengler, J. D., and others. "Personal Exposures to Respirable Particulates and Implications for Air Pollution Epidemiology." *Environmental Science and Technology* 19 (August 1985): 700–707.

Stathos, Dan T., and Michael S. Treitman. "Using Private Market Incentives for Air Cleanup." *Public Utilities Fortnightly* 108 (July 30, 1981): 13–21.

Sweet, William. "Air Pollution Control: Progress and Prospects." *Editorial Research Reports* (November 21, 1980): 843–864.

Tobin, Richard J. "Air Quality and Coal: The U.S. Experience." *Energy Policy* 12 (September 1984): 342–352.

Ulden, A. P. van, and A. A. M. Holtslag. "Estimation of Atmospheric Boundary Layer Parameters for Diffusion Applications." *Journal of Climate and Applied Meteorology* 24 (November 1985): 1196–1207.

Ullmann, Arieh A. "The Implementation of Air Pollution Control in German Industry." *Policy Studies Journal* 11 (September 1982): 141–152.

U.S. Congress, House, Committee on Energy and Commerce, Subcommittee on Health and the Environment. *Clean Air Standards: Hearing*

_____ . *Health Standards for Air Pollutants: Hearings, October 14–15, 1981.* 97th Cong., 1st sess. Washington, DC: GPO, 1982. 499 pp.

_____ . *Toxic Release Control Act of 1985: Hearings, June 11 and 19, 1985 on H.R. 2576, a Bill to Control Toxic Release into the Air, and for Other Purposes.* 99th Cong., 1st sess. Washington, DC: GPO, 1985. 1105 pp.

Wakimoto, R. M. "The Catalina Eddy and Its Effect on Pollution over Southern California." *Monthly Weather Review* 115 (April 1987): 837–855.

Wallace, L. A., and others. "Personal Exposures, Indoor-Outdoor Relationships, and Breath Levels of Toxic Air Pollutants Measured for 355 Persons in New Jersey." *Atmospheric Environment* 19:10 (1985): 1651–1661.

Wanner, H., and J. A. Hertig. "Studies of Urban Climates and Air Pollution in Switzerland." *Journal of Climate and Applied Meteorology* 23 (December 1984): 1614–1625.

Methodology

Statistical Methods

ApSimon, H. M., and A. C. Davison. "A Statistical Model for Deriving Probability Distributions of Contamination for Accidental Releases." *Atmospheric Environment* 20:6 (1986): 1249–1259.

Dana, M. T., and R. C. Easter. "Statistical Summary and Analyses of Event Precipitation Chemistry from the MAP3S Network, 1976–1983." *Atmospheric Environment* 21:1 (1987): 113–128.

Davis, B. L. "X-ray Diffraction Analysis and Source Apportionment of Denver Aerosol." *Atmospheric Environment* 18:10 (1984): 2197–2208.

Henmi, T., and J. F. Bresch. "Meteorological Case Studies of Regional High Sulfur Episodes in the Western United States." *Atmospheric Environment* 19:11 (1985): 1783–1796.

Huber, A. H. "Evaluation of a Method for Establishing Pollution Concentrations Downwind of Influencing Buildings." *Atmospheric Environment* 18:11 (1984): 2313–2338.

Inoue, T., and others. "Regression Analysis of Nitrogen Oxide Concentration [Japan]." *Atmospheric Environment* 20:1 (1986): 71–85.

————. "Stability of Two Prediction Schemes for Hourly Nitrogen Oxide Concentrations by a Regression Model." *Atmospheric Environment* 21:4 (1987): 929–942.

Koning, H. W. de, and others. "Air Pollution in Different Cities Around the World." *Atmospheric Environment* 20:1 (1986): 101–113.

Kuo, Y. H., and others. "The Accuracy of Trajectory Models as Revealed by the Observing System Simulation Experiments." *Monthly Weather Review* 113 (November 1985): 1852–1867.

Liljestrand, H. M. "Average Rainwater pH, Concepts of Atmospheric Acidity, and Buffering in Open Systems." *Atmospheric Environment* 19:3 (1985): 487–499.

McLeod, A. R., and others. "Open-Air Fumigation of Field Crops: Criteria and Design for a New Experimental System." *Atmospheric Environment* 19:10 (1985): 1639–1649.

Noll, K. E., and others. "Comparison of Atmospheric Coarse Particles at an Urban and Non-Urban Site." *Atmospheric Environment* 19:11 (1985): 1931–1943.

Pitchford, M., and A. Pitchford. "Analysis of Regional Visibility in the Southwest Using Principal Component and Back Trajectory Techniques." *Atmospheric Environment* 19:8 (1985): 1301–1316.

Rao, S. T., and others. "Resampling and Extreme Value Statistics in Air Quality Model Performance Evaluation." *Atmospheric Environment* 19:9 (1985): 1503–1518.

Sexton, K., and others. "Characterization and Source Apportionment of Wintertime Aerosol in a Wood-Burning Community." *Atmospheric Environment* 19:8 (1985): 1225–1236.

Streets, D. G., and others. "Targeted Strategies for Control of Acidic Deposition." *Journal of the Air Pollution Control Association* 34 (December 1984): 1187–1197.

Taylor, J. A., and others. "Modeling Distributions of Air Pollutant Concentrations." *Atmospheric Environment* 20:9 (1986); 1781–1789; 20:12 (1986): 2435–2447.

Thurston, G. D., and P. J. Lioy. "Receptor Modeling and Aerosol Transport." *Atmospheric Environment* 21:3 (1987): 687–698.

Wendling, P., and others. "Calculated Radiative Effects of Arctic Haze During a Pollution Episode in Spring 1983 Based on Ground-Based and Airborne Measurements." *Atmospheric Environment* 19:12 (1985): 2181–2193.

Wilson, D. J., and others. "Intermittency and Conditionally-Averaged Concentration Fluctuation Statistics in Plumes." *Atmospheric Environment* 19:7 (1985): 1053–1064.

Wiman, B.L.B., and H. O. Lannefors. "Aerosol Characteristics in a Mature Coniferous Forest—Methodology, Composition, Sources, and Spatial Concentration Variations." *Atmospheric Environment* 19:2 (1985): 349–362; Discussion 20:2 (1986): 407–408.

̅̅̅̅̅, ̅. ̅., and others. "Measurement of Atmospheric Aerosols and Photochemical Products at a Rural Site in SW Ontario." *Atmospheric Environment* 19:11 (1985): 1859–1870.

Businger, J. A. "Evaluation of the Accuracy with Which Dry Deposition Can Be Measured with Current Micrometeorological Techniques." *Journal of Climate and Applied Meteorology* 25 (August 1986): 1100–1124.

Isaac, G. A., and others. "Summer Aerosol Profiles over Algonquin Park, Canada." *Atmospheric Environment* 20:1 (1986): 157–172.

Johnson, D. L., and others. "Chemical and Physical Analyses of Houston Aerosol for Interlaboratory Comparison of Source Apportionment Procedures." *Atmospheric Environment* 18:8 (1984): 1539–1553.

Lewin, E. E., and others. "Atmospheric Gas and Particle Measurements at a Rural Northeastern U.S. Site [Pennsylvania]." *Atmospheric Environment* 20:1 (1986): 59–70.

Oblad, M., and E. Selin. "Measurements of Elemental Composition in Background Aerosol on the West Coast of Sweden." *Atmospheric Environment* 20:7 (1986): 1419–1432.

Ottar, B., and others. "Aircraft Measurements of Air Pollution in the Norwegian Arctic." *Atmospheric Environment* 20:1 (1986): 87–100.

Tang, A.S.J., and others. "An Analysis of the Impact of the Sudbury Smelters on Wet and Dry Deposition in Ontario." *Atmospheric Environment* 21:4 (1987): 813–824.

Testing

Cheremisinoff, P. N. "Special Report on Air Toxics: Measuring and Monitoring." *Pollution Engineering* 17 (June 1985): 21–29.

Eskridge, R. E., and S. Trivikrama Rao. "Measurement and Prediction of Traffic-Induced Turbulence and Velocity Fields near Roadways." *Journal of Climate and Applied Meteorology* 22 (August 1983): 1431–1443; Discussion 24 (September 1985): 1003–1007.

Hard, T. M., and others. "Tropospheric Free Radical Determination by FAGE [Fluorescence Assay with Gas Expansion]." *Environmental Science and Technology* 18 (October 1984): 768–777.

Illes, S. P. "A Portable Data-Logging System for Industrial Hygiene Personal Chlorine Monitoring." *American Industrial Hygiene Association Journal* 47 (February 1986): 78–86.

Lewin, E. E., and others. "Atmospheric Gas and Particle Measurements at a Rural Northeastern U.S. Site." *Atmospheric Environment* 20:1 (1986): 59–70.

Nyren, K., and S. Winter. "Discharge of Condensed Sulfur Dioxide: A Field Test Study of the Source Behaviour with Different Release Geometries." *Journal of Hazardous Materials* 14 (March 1987): 365–386.

Oppenheimer, M., and others. "Acid Deposition, Smelter Emissions, and the Linearity Issue in the Western United States." *Science* 229 (August 30, 1985): 859–862; Discussion 233 (July 4, 1986): 10–14.

Pueschel, R. F., and others. "Aerosols in Polluted Versus Nonpolluted Air Masses: Long-Range Transport and Effects on Clouds." *Journal of Climate and Applied Meteorology* 25 (December 1986): 1908–1917.

Rodes, C. E., and E. G. Evans. "Preliminary Assessment of 10 um Particulate Sampling at Eight Locations in the United States." *Atmospheric Environment* 19:2 (1985): 293–303.

Thrane, K. E. "Ambient Air Concentrations of Polycyclic Aromatic Hydrocarbons, Fluoride, Suspended Particles, and Particulate Carbon in Areas near Aluminum Production Plants." *Atmospheric Environment* 21:3 (1987): 617–628.

Wadden, R. A. "Source Discrimination of Short-Term Hydrocarbon Samples Measured Aloft." *Environmental Science and Technology* 20 (May 1986): 473–483.

Standards

Davidson, J. E., and P. K. Hopke. "Implications of Incomplete Sampling on a Statistical Form of the Ambient Air Quality Standard for Particulate Matter." *Environmental Science and Technology* 18 (August 1984): 571–580.

Larsen, R. I., and W. W. Heck. "Air Quality Data Analysis System for Interrelating Effects, Standards, and Needed Source Reductions." *Journal of the Air Pollution Control Association* 33 (March 1983): 198–207; 34 (October 1984): 1023–1034; 35 (December 1985): 1274–1279.

U.S. Congress, House, Committee on Energy and Commerce, Subcommittee on Oversight and Investigations. *Air Quality Standards: Hearing, October 1, 1984.* 98th Cong., 2d sess. Washington, DC: GPO, 1985. 1467 pp.

Mathematical Models

Austin, L. S., and G. E. Millward. "Modelling Temporal Variations in the Global Tropospheric Arsenic Burden." *Atmospheric Environment* 18:9 (1984): 1909–1919.

Daisey, J. M., and others. "Profiles of Organic Particulate Emissions from Air Pollution Sources: Status and Needs for Receptor Source Apportionment Modeling." *Journal of the Air Pollution Control Association* 36 (January 1986): 17–33.

Dennis, R. L., and M. W. Downton. "Evaluation of Urban Photochemical Models for Regulatory Use [Denver, CO]." *Atmospheric Environment* 18:10 (1984): 2055–2069.

Ell. . . . . *Atmospheric Environment* 19:5 (1985): 727–737.

Ellis, J. H., and others. "Deterministic Linear Programming Model for Acid Rain Abatement." *Journal of Environmental Engineering* 111 (April 1985): 119–139.

Endlich, R. M., and others. "A Long-Range Air Pollution Transport Model for Eastern North America [ENAMAP]." *Atmospheric Environment* 18:1 (1984): 2345–2366.

Hildemann, L. M., and others. "Ammonia and Nitric Acid Concentrations in Equilibrium with Atmospheric Aerosols: Experiment vs. Theory [California]." *Atmospheric Environment* 18:9 (1984): 1737–1750.

Karamchandani, P., and L. K. Peters. "Three-Dimensional Behavior of Mixing-Limited Chemistry in the Atmosphere." *Atmospheric Environment* 21:3 (1987): 511–522.

Khalil, M.A.K., and R. A. Rasmussen. "Causes of Increasing Atmospheric Methane: Depletion of Hydroxyl Radicals and the Rise of Emissions." *Atmospheric Environment* 19:3 (1985): 397–407.

Kitada, T., and others. "Numerical Simulation of the Transport of Chemically Reactive Species Under Land- and Sea-Breeze Circulations." *Journal of Climate and Applied Meteorology* 23 (August 1984): 1153–1172.

Ku, J. Y., and others. "Numerical Simulation of Air Pollution in Urban Areas: Model Development [and Model Performance]." *Atmospheric Environment* 21:1 (1987): 201–232.

Kumar, Sudarshan. "Reactive Scavenging of Pollutants by Rain: A Modeling Approach." *Atmospheric Environment* 20:5 (1986): 1015–1024.

Pilinis, C., and others. "Mathematical Modeling of the Dynamics of Multicomponent Atmospheric Aerosols [Los Angeles, CA]." *Atmospheric Environment* 21:4 (1987): 943–955.

Russell, A. G., and G. R. Cass. "Acquisition of Regional Air Quality Model Validation Data for Nitrate Sulfate, Ammonium Ion, and Their Precursors [California]." *Atmospheric Environment* 18:9 (1984): 1815–1827.

Sloane, C. S., and G. T. Wolff. "Prediction of Ambient Light Scattering Using a Physical Model Responsive to Relative Humidity: Validation with Measurements from Detroit." *Atmospheric Environment* 19:4 (1985): 669–680.

Tremblay, A., and H. Leighton. "The Influence of Cloud Dynamics upon the Redistribution and Transformation of Atmospheric SO_2—A Numerical Simulation." *Atmospheric Environment* 18:9 (1984): 1885–1894.

Wilczak. J. M., and M. S. Phillips. "An Indirect Estimation of Convective Boundary Layer Structure for Use in Pollution Dispersion Models." *Journal of Climate and Applied Meteorology* 25 (November 1986): 1609–1624.

Wiman, B.L.B., and G. I. Agren. "Aerosol Depletion and Deposition in Forests—A Model Analysis." *Atmospheric Environment* 19:2 (1985): 335–347.

Yadigaroglu, G., and H. A. Munera. "Transport of Pollutants: Summary Review of Physical Dispersion Models." *Nuclear Technology* 77 (May 1977): 125–149.

Yamartino, R. J., and G. Wiegand. "Development and Evaluation of Simple Models for the Flow, Turbulence, and Pollutant Concentration Fields Within an Urban Street Canyon [Cologne, W. Germany]." *Atmospheric Environment* 20:11 (1986): 2137–2156.

Research

Altshuller, A. P. "The Role of Nitrogen Oxides in Nonurban Ozone Formation in the Planetary Boundary Layer over N. America, W. Europe, and Adjacent Areas of the Ocean." *Atmospheric Environment* 20:2 (1986): 245–268.

Atkinson, R. "Kinetics and Mechanisms of the Gas-Phase Reactions of the Hydroxyl Radical with Organic Compounds Under Atmospheric Conditions." *Chemical Reviews* 86 (February 1986): 69–201.

Krogstad, P. A., and R. M. Pettersen. "Windtunnel Modelling of a Release of Heavy Gas Near a Building." *Atmospheric Environment* 20:5 (1986): 867–878.

Mansfeld, F., and others. "A New Atmospheric Corrosion Rate Monitor—Development and Evaluation." *Atmospheric Environment* 20:6 (1986): 1179–1192.

Mazurek, M. A., and B.R.T. Simoneit. "Organic Components in Bulk and Wet-Only Precipitation." *Critical Reviews in Environmental Control* 16:1 (1986): 1–140.

Orgill, M. M., and R. I. Schreck. "An Overview of the ASCOT Multi-Laboratory Field Experiments in Relation to Drainage Winds and Ambient Flow." *Bulletin of the American Meteorological Society* 66 (October 1985): 1263–1277.

Stockwell, W. R. "A Homogeneous Gas Phase Mechanism for Use in a Regional Acid Deposition Model." *Atmospheric Environment* 20:8 (1986): 1615-1632.

Particulate Pollution
General

Cowherd, Chatten, Jr., and Phillip J. Englehart. *Size Specific Particulate Emission Factors* ~~~ ... ~~~ ..., ... U.S. Environmental Protection Agency, Air and Energy Engineering Research Laboratory, 1986. 3 pp.

Dzubay, T. G., and others. "Interlaboratory Comparison of Receptor Model Results for Houston Aerosol." *Atmospheric Environment* 18:8 (1984): 1555-1566.

Fukuzaki, N., and others. "Effects of Studded Tires on Roadside Airborne Dust Pollution in Niigata, Japan." *Atmospheric Environment* 20:2 (1986): 377-386.

Gray, H. A., and others. "Characteristics of Atmospheric Organic and Elemental Carbon Particle Concentrations in Los Angeles." *Environmental Science and Technology* 20 (June 1986): 580-589.

Greenberg, A., and others. "Polycyclic Aromatic Hydrocarbons in New Jersey: A Comparison of Winter and Summer Concentrations over a Two-Year Period." *Atmospheric Environment* 19:8 (1985): 1325-1339.

Hobbs, P. V., and J. C. Yates. "Atmospheric Aerosol Measurements over North America and the North Atlantic Ocean." *Atmospheric Environment* 19:1 (1985): 163-179.

Lowenthal, D. H., and K. A. Rahn. "Regional Sources of Pollution Aerosol at Barrow, Alaska, During Winter 1979-80, as Deduced from Elemental Tracers." *Atmospheric Environment* 19:12 (1985): 2011-2024.

Mamane, Y., and K. E. Noll. "Characterization of Large Particles at a Rural Site in the Eastern United States: Mass Distribution and Individual Particle Analysis." *Atmospheric Environment* 19:4 (1985): 611-622.

Marshall, B. T., and others. "Characterization of the Atlanta Area Aerosol: Elemental Composition of Possible Sources." *Atmospheric Environment* 20:6 (1986): 1291-1300.

Martinson, B. G., and others. "Southern Scandinavian Aerosol Composition and Elemental Size Distribution Characteristics Dependence on Air Mass History." *Atmospheric Environment* 18:10 (1984): 2167-2182.

Noll, K. E., and others. "Comparison of Atmospheric Coarse Particles at an Urban and Non-Urban Site [Illinois]." *Atmospheric Environment* 19:11 (1985): 1931-1943.

Okada, K. "Number-Size Distribution and Formation Process of Sub-micrometer Sulfate-Containing Particles in the Urban Atmosphere of Nagoya." *Atmospheric Environment* 19:5 (1985): 743–757.

Orsini, C. Q., and others. "Characteristics of Fine and Coarse Particles of Natural and Urban Aerosols of Brazil." *Atmospheric Environment* 20:11 (1986): 2259–2269.

Poirot, R. L., and P. R. Wishinski. "Visibility Sulfate and Air Mass History Associated with the Summertime Aerosol in Northern Vermont." *Atmospheric Environment* 20:7 (1986): 1457–1469.

Shaw, G. E. "Aerosol Measurements in Central Alaska, 1982–1984." *Atmospheric Environment* 19:12 (1985): 2025–2031.

Sturges, W. T., and R. M. Harrison. "Bromine in Marine Aerosols and the Origin, Nature, and Quantity of Natural Atmospheric Bromine [Great Britain]." *Atmospheric Environment* 20:7 (1986): 1485–1496.

————. "The Use of Br/Pb Ratios in Atmospheric Particles to Discriminate Between Vehicular and Industrial Lead Sources in the Vicinity of a Lead Works: Thorpe, West Yorkshire [and Ellesmere Port, Cheshire]." *Atmospheric Environment* 20:5 (1986): 833–850; Correction 21:4 (1987): 1005.

Thurston, G. D., and J. D. Spengler. "A Quantitative Assessment of Source Contributions to Inhalable Particulate Matter Pollution in Metropolitan Boston." *Atmospheric Environment* 19:1 (1985): 9–25; Discussion 21:1 (1987): 257–260.

VanVaeck, L., and K. A. Van Cauwenberghe. "Characteristic Parameters of Particle Size Distributions of Primary Organic Constituents of Ambient Aerosols [Belgium]." *Environmental Science and Technology* 19 (August 1985): 707–716.

Vossler, T. L., and E. S. Macias. "Contribution of Fine Particle Sulfates to Light Scattering in St. Louis Summer Aerosol." *Environmental Science and Technology* 20 (December 1986): 1235–1243.

Wolff, G. T., and P. E. Korsog. "Estimates of the Contributions of Sources to Inhalable Particulate Concentrations in Detroit." *Atmospheric Environment* 19:9 (1985): 1399–1409.

Wolff, G. T., and others. "The Influence of Local and Regional Sources on the Concentration of Inhalable Particulate Matter in Southeastern Michigan." *Atmospheric Environment* 19:2 (1985): 305–313.

Fires

Chung, Y. S. "On the Forest Fires and the Analysis of Air Quality Data and Total Atmospheric Ozone." *Atmospheric Environment* 18:10 (1984): 2153–2157.

Lipfert, F. W., and J. Lee. "Air Pollution Implications of Increasing Residential Firewood Use." *Energy* 10 (January 1985): 17–33.

"NASA to Study Effect of Forest Fires on Global Air Quality." *Journal of the Air Pollution Control Association* 36 (November 1986): 1291–1293.

Roper, William. "Wood Stoves: Can We Solve the Emissions Problem Before It Goes Up in Smoke?" *Boston College Environmental Affairs Law Review* 11 (January 1984): 273–294.

of a Large Forest Fire. *Applied Optics* 25 (August 1, 1986): 2554–2562.

Volcanic Dust

Angell, J. K., and J. Korshover. "Surface Temperature Changes Following the Six Major Volcanic Episodes Between 1780 and 1980." *Journal of Climate and Applied Meteorology* 24 (September 1985): 937–951; Discussion 25 (August 1986): 1184–1186.

Rampino, M. R., and S. Self. "The Atmospheric Effects of El Chichón." *Scientific American* 250 (January 1984): 48–57.

Schneider, S. H., and C. Moss. "Volcanic Dust, Sunspots, and Temperature Trends." *Science* 190 (November 21, 1975): 741–746.

Woodcock. A. H. "Mountain Breathing Revisited—The Hyperventilation of a Volcano Cinder Cone." *Bulletin of the American Meteorological Society* 68 (February 1987): 125–130.

Woods, D. C., and others. "Halite Particles Injected into the Stratosphere by the 1982 El Chichón Eruption." *Science* 230 (October 11, 1985): 170–172.

Acid Precipitation

"Acid Rain: A Major Threat to the Ecosystem." *Conservation Foundation Letter* (December 1982): 1–8.

"Acid Rain: Looking Ahead." *EPA (Environmental Protection Agency) Journal* 12 (June–July 1986): 2–26.

"Acid Rain Options." *Journal of the Air Pollution Control Association* 35 (March 1985): 197–226.

Adams, R. M., and others. "Pollution, Agriculture, and Social Welfare: The Case of Acid Deposition." *Canadian Journal of Agricultural Economics* 34 (March 1986): 3–19.

Audette, Rose Marie L. "Acid Rain Is Killing More Than Lakes and Trees." *Environmental Action* 18 (May–June 1987): 10–13.

Barrie, L. A., and R. M. Hoff. "The Oxidation Rate and Residence Time of Sulphur Dioxide in the Arctic Atmosphere." *Atmospheric Environment* 18:12 (1984): 2711–2722.

Bilonick, R. A. "The Space-Time Distribution of Sulfate Deposition in the Northeastern United States." *Atmospheric Environment* 19:11 (1985): 1829–1845.

Cowling, Ellis B. "What Is Happening to Germany's Forests? 'Scientifically Unprecedented Changes.'" *Environmental Forum* 3 (May 1984): 6–11.

Davidson, C. I., and others. "The Scavenging of Atmospheric Sulfate by Arctic Snow." *Atmospheric Environment* 21:4 (1987): 871–882.

Dzubay, T. G., and others. "Composition and Origins of Aerosol at a Forested Mountain in Soviet Georgia." *Environmental Science and Technology* 18 (November 1984): 873–883.

Feddema, J. J., and T. C. Meierding. "Marble Weathering and Air Pollution in Philadelphia." *Atmospheric Environment* 21:1 (1987): 143–157.

Gaffney, J. S., and others. "Beyond Acid Rain: Do Soluble Oxidants and Organic Toxins Interact with SO_2 and NO_2 to Increase Ecosystem Effects?" *Environmental Science and Technology* 21 (June 1987): 519–524.

Hegg, D. A., and others. "Field Studies of a Power Plant Plume in the Arid Southwestern United States." *Atmospheric Environment* 19:7 (1985): 1147–1167.

Henmi, T., and J. F. Bresch. "Meteorological Case Studies of Regional High Sulfur Episodes in the Western United States." *Atmospheric Environment* 19:11 (1985): 1783–1796.

Hicks, B. B., and others. *On the Use of Air Concentrations to Infer Dry Deposition.* NOAA Technical Memorandum ERL ARL, 141. Silver Spring, MD: National Oceanic and Atmospheric Administration, Environmental Research Laboratories, Air Resources Laboratory, 1985. 65 pp.

Hidy, G. M. "Source-Receptor Relationships for Acid Deposition: Pure and Simple? [North America]." *Journal of the Air Pollution Control Association* 34 (May 1984): 518–531; Dicussion 34 (September 1984): 905–919; 35 (February 1985): 127–134.

Krug, E. C., and C. R. Frank. "Acid Rain on Acid Soil: A New Perspective." *Science* 225 (September 28, 1984): 1424–1434; Discussion 221 (August 5, 1983): 520–525.

Pechman, Carl. "Equity, Efficiency, and Sulfur Emission Reduction." *Public Utilities Fortnightly* 15 (May 16, 1985): 21–28.

Pierson, W. R., and T. Y. Chang. "Acid Rain in Western Europe and Northeastern United States—A Technical Appraisal." *Critical Reviews in Environmental Control* 16:2 (1986): 167–192.

Regens, James L. "The Political Economy of Acid Rain." *Publius* 15 (Summer 1985): 53–66.

Regens, James L., and Robert W. Rycroft. "Options for Financing Acid Rain Controls." *Natural Resources Journal* 26 (Summer 1986): 519–549.

Sisterson, D. L., and others. "Chemical Differences Between Event and Weekly Precipitation Samples in Northeastern Illinois." *Atmospheric Environment* 19:9 (1985): 1453–1469.

Special Issue: Acid Rain I

Sweet, William. "Acid Rain." *Editorial Research Reports* (June 20, 1980): 447–464.

Tabatabai, M. A. "Effect of Acid Rain on Soils." *Critical Reviews in Environmental Control* 15:1 (1985): 65–110.

Tonnessen, Kathy, and John Harte. "Acid Rain and Ecological Damage: Implications of Sierra Nevada Lake Studies." *Public Affairs Report* (December 1982): 1–9.

Trisko, Eugene M. "Acid Precipitation: Causes, Consequences, Controls." *Public Utilities Fortnightly* 111 (February 3, 1983): 19–27.

U.S. Congress, House. *Impact of Air Pollutants on Agriculture Productivity: Joint Hearing, July 9, 1981, Before the Subcommittee on Natural Resources, Agriculture Research, and Environment of the Committee on Science and Technology and the Subcommittee on Department Operations, Research, and Foreign Agriculture of the Committee on Agriculture.* 97th Cong., 1st sess. Washington, DC: GPO, 1981. 148 pp.

U.S. Congress, Senate, Committee on Energy and Natural Resources. *Clean Coal Technology Development and Strategies for Acid Rain Control: Hearings, June 9 and 10, 1986.* 99th Cong. 2d sess. Washington, DC: GPO, 1987. 935 pp.

Young, W. S., and R. W. Shaw. "A Proposed Strategy for Reducing Sulphate Deposition in North America: Methodology for Minimizing Sulphur Removal and Methodology for Minimizing Costs." *Atmospheric Environment* 20:1 (1986): 189–206.

Carbon Dioxide

Bryson, R. A. "Climatic Modification by Air Pollution." In Polunin, N., ed., *The Environmental Future*, pp. 134–154. London: Macmillan, 1972.

_____. "A Perspective on Climatic Change." *Science* 184 (May 17, 1974): 753–760.

Cess, R. D. "Climate Change: An Appraisal of Atmospheric Feedback Mechanisms Employing Zonal Climatology." *Journal of the Atmospheric Sciences* 33:11 (1976): 1831–1843.

Idso, S. B. "Carbon Dioxide and Climate: Is There a Greenhouse in Our Future?" *Quarterly Review of Biology* 59 (September 1984): 291–294.

————. "Carbon Dioxide and Global Temperature: What the Data Show." *Journal of Environmental Quality* 12 (April–June 1983): 159–163.

————. "Implications of the CO_2 Greenhouse Effect for Agriculture and Water Resources: A Critical Review of Recent Official Reports of the U.S. National Research Council and the U.S. Environmental Protection Agency." *Soil Science* 138 (November 1984): 373–383.

Jäger, J. "Climatic Change: Floating New Evidence in the CO_2 Debate." *Environment* 28 (September 1986): 6–9+.

Lamb, H. H. "Volcanic Dust in the Atmosphere; with a Chronology and Assessment of Its Meteorological Significance." *Philosophical Transactions of the Royal Society of London: Mathematical and Physical Sciences* 266:1178 (1970): 425–533.

Nanda, Ved P. "Global Climate Change and International Law." *Impact of Science on Society* 32 (July–September 1982): 365–374.

Peters, R. L., and J.D.S. Darling. "The Greenhouse Effect and Nature Reserves." *BioScience* 35 (December 1985): 707–717.

Pittock, A. B. "Report on Reports: The Carbon Dioxide Debate: Reports from SCOPE and DOE." *Environment* 29 (January–February 1987): 25–30.

Schelling, T. C. "Anticipating Climate Change: Implications for Welfare and Policy." *Environment* 26 (October 1984): 6–9+.

"Symposium: Global Climatic Change." Papers presented at a conference, Denver, CO, July 10–11, 1980. *Denver Journal of International Law and Policy* 10 (Spring 1981): 463–535.

U.S. Congress, House, Committee on Science and Technology. *Carbon Dioxide and Climate: The Greenhouse Effect: Hearing, July 31, 1981, Before the Subcommittee on Natural Resources, Agriculture Research, and Environment and the Subcommittee on Investigations and Oversight.* 97th Cong., 1st sess. Washington, DC: GPO, 1981. 148 pp.

Winner, W. E., and T. J. Casadevall. "The Effects of the Mount St. Helens Eruption Cloud on Fir (Abies sp) Needle Cuticles: Analysis with Scanning Electron Microscopy." *American Journal of Botany* 70 (January 1983): 80–87.

Woodwell, George M. "The Carbon Dioxide Question." *Scientific American* 238 (January 1978): 34–43.

Ozone

Altshuller, A. P. "Relationships Between Direction of Wind Flow and Ozone Inflow Concentrations at Rural Locations Outside of St. Louis, MO." *Atmospheric Environment* 20:11 (1986): 2175–2184.

Angle, R. P., and H. S. Sandhu. "Rural Ozone Concentrations in Alberta, Canada." *Atmospheric Environment* 20:6 (1986): 1221–1228.

Broder, B., and H. A. Gygax. "The Influence of Locally Induced Wind Systems on the Effectiveness of Nocturnal Dry Deposition of Ozone [Switzerland]." *Atmospheric Environment* 19:11 (1985): 1627–1637.

Camm, F., and others. *Social Cost of Technical Control Options to Reduce Emissions of Potential Ozone Depleters in the United States. An Update*

[...]sook, I., and R. M. Harrison. The Frequency and Causes of Elevated Concentrations of Ozone at Ground Level at Rural Sites in North-West England." *Atmospheric Environment* 19:10 (1985): 1577–1587.

Lefohn, A. S., and others. "The Development of Sulfur Dioxide and Ozone Exposure Profiles That Mimic Ambient Conditions in the Rural Southeastern United States." *Atmospheric Environment* 21:3 (1987): 659–669.

Meagher, J. F. "Rural Ozone in the Southeastern United States." *Atmospheric Environment* 21:3 (1987): 605–615.

Mehlman, M. A., and C. Borek. "Toxicity and Biochemical Mechanisms of Ozone." *Environmental Research* 42 (February 1987): 36–53.

Merikaarto, Kaarina. "Ozone Away." *Environmental Action* 18 (March-April 1987): 14–18.

U.S. Congress, House, Committee on Energy and Commerce, Subcommittee on Oversight and Investigations. *EPA: Ozone and the Clean Air Act: Hearing, April 27, 1987.* 100th Cong., 1st sess. Washington, DC: GPO, 1987. 1650 pp.

Zellner, K., and N. Moussiopoulos. "Simulations of the Ozone Formations Caused by Traffic in Urban Areas." *Atmospheric Environment* 20:8 (1986): 1589–1596.

Radioactive Contamination

General

Bopp, Anthony, and others. "Air Quality Implications of a Nuclear Moratorium: An Alternative Analysis." *Energy Journal* 2 (July 1981): 33–48.

Colbeck, I., and R. M. Morrison. "The Atmospheric Effects of Nuclear War—A Review." *Atmospheric Environment* 20:9 (1986): 1673–1681.

Gupta, V. K., and R. K. Kapoor. "Reducing the Consequences of Reactor Accidents with a Green Belt." *Nuclear Technology* 70 (August 1985): 204–214.

Stensvaag, John-Mark. "Regulating Radioactive Air Emissions from Nuclear Generating Plants: A Primer for Attorneys, Decisionmakers, and Intervenors." *Northwestern University Law Review* 78 (March 1983): 1-197.

Chernobyl Accident

ApSimon, H., and J. Wilson. "Tracking the Cloud from Chernobyl." *New Scientist* 111 (July 17, 1986): 42-45.

Fischetti, M. A. "The Puzzle of Chernobyl." *IEEE Spectrum* 23 (July 1986): 34-41.

Franklin, N. "The Accident at Chernobyl." *Chemical Engineer* (England) no. 430 (November 1986): 17-22.

Hohenemser, C., and others. "Chernobyl." *Chemtech* 16 (October 1986): 596-605.

Kauppinen, E. I., and others. "Radioactivity Size Distributions of Ambient Aerosols in Helsinki, Finland, During May 1986 After Chernobyl Accident: Preliminary Report." *Environmental Science and Technology* 20 (December 1986): 1257-1259.

Kolata, G. "The UCLA-Occidental-Gorbachev Connection." *Science* 233 (July 4, 1986): 19-21.

Levi, B. G. "Cause and Impact of Chernobyl Accident Still Hazy." *Physics Today* 39 (July 1986): 17-21.

Malinauskas, A. P., and others. "Calamity at Chernobyl." *Mechanical Engineering* 109 (February 1987): 50-53: Discussion 109 (May 1987): 4.

Rossi, P. H., and others. "The Urban Homeless: Estimating Composition and Size." *Science* 235 (March 13, 1987): 1336-1341.

Simmonds, J. "Europe Calculates the Health Risk." *New Scientist* 114 (April 23, 1987): 40-43.

U.S. Congress, House, Committee on Science and Technology, Subcommittee on Energy Research and Production. *Positive Safety Features of U.S. Nuclear Reactors; Technical Lessons Confirmed at Chernobyl: Hearing, May 14, 1986.* 99th Cong., 2d sess. Washington, DC: GPO, 1986. 125 pp.

U.S. Congress, Joint Economic Committee, Subcommittee on Agriculture and Transportation. *The Chernobyl Disaster; Implications for World Food Security and the U.S. Farm Economy: Hearing Before the Subcommittee on Agriculture and Transportation of the Joint Economic Committee.* 99th Cong., 2d sess. Washington, DC: GPO, 1987. 68 pp.

U.S. Congress, Senate, Committee on Agriculture, Nutrition, and Forestry. *Possible Impact on Agriculture of the Explosion of the Soviet Nuclear Plant at Chernobyl: Hearing, May 15, 1986.* 99th Cong., 2d sess. Washington, DC: GPO, 1986. 59 pp.

U.S. Congress, Senate, Committee on Energy and Natural Resources. *The Chernobyl Accident: Hearing, June 19, 1986, Before the Committee on Energy and National Resources on the Chernobyl Accident and Implications for the Domestic Nuclear Industry.* 99th Cong., 2d sess. Washington, DC: GPO, 1986. 343 pp.

U.S. Congress, Senate, Committee on Governmental Affairs, Subcommittee on Energy, Nuclear Proliferation, and Government Processes. *International Nuclear Safety: C̲____ __ H____ ___*

USSR State Conference on the Utilization of Atomic Energy, August 25–29, 1986, Vienna, Austria. *The Accident at the Chernobyl Nuclear Power Plant and Its Consequences.* Information Compiled for the IAEA Experts Meeting: USSR State Committee on the Utilization of Atomic Energy. Vienna, Austria: International Atomic Energy Agency, 1986. 3 vols.

Webster, D. "How Ministers Misled Britain About Chernobyl." *New Scientist* 112 (October 9, 1986): 43–46.

"World News—Chernobyl." *Nuclear Engineering International* 31 (October 1986): 2–9.

Photochemical Air Pollution

Smog

Albersheim, Steven R. "An Assessment of Transportation Control Measures for Improving Air Quality." *Transportation Quarterly* 36 (July 1982): 451–468.

Evans, L. F., and others. "A Chamber Study of Photochemical Smog in Melbourne, Australia—Present and Future." *Atmospheric Environment* 20:7 (1986): 1355–1368.

Fox, William F., Jr. "Getting the Lead Out Of American Gasoline." *Journal of Energy and Natural Resources Law* 4:1 (1986): 1–9.

Israel, G. W., and others. "The Berlin Smog Project—Description and Summary of Results." *Atmospheric Environment* 18:10 (1984): 2071–2088.

Kurita, H., and H. Veda. "Meteorological Conditions for Long-Range Transport Under Light Gradient Winds." *Atmospheric Environment* 20:4 (1986): 687–694.

Marsden, A. R., Jr., and others. "Increasing the Computational Feasibility of Urban Air Quality Models That Employ Complex Chemical Mechanisms." *JAPCA* 37 (April 1987): 370–376.

Matsoukis, E. C. "An Assessment of Vehicle Restraint Measures [Greece]." *Transportation Quarterly* 39 (January 1985): 125–133.

Newcomb, Tim M. "The Seattle Automobile Inspection and Maintenance Program: Multiple Analyses of Program Impact." *Evaluation Review* 10 (April 1986): 217–229.

Newman, P.W.G., and J. R. Kenworthy. "The Use and Abuse of Driving Cycle Research: Clarifying the Relationship Between Traffic Congestion, Energy, and Emissions." *Transportation Quarterly* 38 (October 1984): 615–635.

Ostrov, Jerome. "Inspection and Maintenance of Automotive Pollution Controls: A Decade-Long Struggle Among Congress, EPA [Environmental Protection Agency], and the States." *Harvard Environmental Law Review* 8:1 (1984): 139–191.

Richardson, James. "The Politics of Smog." *California Journal* 18 (June 1987): 284–289.

Rowe, M. D. "NO_2 Exposure from Vehicles and Gas Stoves." *Journal of Transportation Engineering* 111 (November 1985): 679–691.

Sigsby, J. E., Jr., and others. "Volatile Organic Compound Emissions from 46 In-Use Passenger Cars." *Environmental Science and Technology* 21 (March 1987): 466–475.

U.S. Congress, House, Committee on Government Operations, Subcommittee on Environment, Energy, and Natural Resources. *Lead in Gasoline; Public Health Dangers: Hearing, April 14, 1982.* 97th Cong., 2d sess. Washington, DC: GPO, 1982. 307 pp.

U.S. Congress, House, Committee on Science and Technology, Subcommittee on Investigations and Oversight. *Diesel Technology: Hearing, May 6, 1982.* 97th Cong., 2d sess. Washington, DC: GPO, 1983. 190 pp.

Uno, I., and others. "Evaluation of Hydrocarbon Reactivity in Urban Air [Tokyo, Japan]." *Atmospheric Environment* 19:8 (1985): 1283–1293.

———. "Three-Dimensional Behaviour of Photochemical Pollutants Covering the Tokyo Metropolitan Area." *Atmospheric Environment* 18:4 (1984): 751–761.

Wakimoto, R. M., and J. L. McElroy. "Lidar Observation of Elevated Pollution Layers over Los Angeles." *Journal of Climate and Applied Meteorology* 25 (November 1986): 1583–1599.

Haze

Chang, W. L., and H. E. Koo. "A Study of Visibility Trends in Hong Kong." *Atmospheric Environment* 20:10 (1986): 1847–1858.

Kaufman, Y. J., and T. W. Brakke. "Field Experiment for Measurement of the Radiative Characteristics of a Hazy Atmosphere." *Journal of the Atmospheric Sciences* 43 (June 1, 1986): 1135–1151.

Lewis, C. W., and others. "Receptor Modeling Study of Denver Winter Haze." *Environmental Science and Technology* 20 (November 1986): 1126–1136.

Raatz, W. E. "Meteorological Conditions over Eurasia and the Arctic Contributing to the March 1983 Arctic Haze Episode." *Atmospheric Environment* 19:12 (1985): 2121–2126.

Raatz, W. E., and others. "Air M~~~ Cl~~~ ~~~~~~~~~ ~~~~~~~~~ ~~~~~~~ 19:12 (~~~~), ~~~~–~~~~.

_____ . "The Distribution and Transport of Pollution Aerosols over the Norwegian Arctic on 31 March and 4 April 1983." *Atmospheric Environment* 19:12 (1985): 2135–2142.

_____ . "Observations of Arctic Haze During Polar Flights from Alaska to Norway." *Atmospheric Environment* 19:12 (1985): 2143–2151.

Richards, L. W., and others. "The Optical Effects of Fine-Particle Carbon on Urban Atmosphere." *Atmospheric Environment* 20:2 (1986): 387–396.

Sheridan, P. J., and I. H. Musselman. "Characterization of Aircraft-Collected Particles Present in the Arctic Aerosol; Alaskan Arctic, Spring, 1983." *Atmospheric Environment* 19:12 (1985): 2159–2166.

Solomon, P. A., and J. L. Moyers. "A Chemical Characterization of Wintertime Haze in Phoenix, Arizona." *Atmospheric Environment* 20:1 (1986): 207–213.

Vinzani, P. G., and P. J. Lamb. "Temporal and Spatial Visibility Variations in the Illinois Vicinity During 1949–80." *Journal of Climate and Applied Meteorology* 24 (May 1985): 435–451.

Wolff, G. T., and others. "Measurements of Sulfur Oxides, Nitrogen Oxides, Haze, and Fine Particles at a Rural Site on the Atlantic Coast." *Journal of the Air Pollution Control Association* 36 (May 1986): 585–591.

Indoor Pollution

General

Greenberg, Michael R. "Indoor Air Quality: Protecting Public Health Through Design, Planning, and Research." *Journal of Agricultural and Planning Research* 3 (August 1986): 253–261.

Kirsch, Laurence S. "Behind Closed Doors: Indoor Air Pollution and Government Policy." *Harvard Environmental Law Review* 6:2 (1982): 339–394.

Nero, Anthony V., Jr. "Controlling Indoor Air Pollution." *Scientific American* 258 (May 1988): 42–49.

Sexton, Ken. "Indoor Air Quality: An Overview of Policy and Regulatory Issues." *Science, Technology, and Human Values* 11 (Winter 1986): 53–67.

Sexton, Ken, and S. B. Hayward. "Source Apportionment of Indoor Air Pollution." *Atmospheric Environment* 21:2 (1987): 407–418.

U.S. Congress, House, Committee on Science and Technology. *Indoor Air Quality Research: Hearings, August 2–3, 1983, Before the Subcommittee on Energy Development and Applications and the Subcommittee on Natural Resources, Agriculture Research, and Environment.* 98th Cong., 1st sess. Washington, DC: GPO, 1984. 508 pp.

U.S. Congress, Senate, Committee on Environment and Public Works. *Indoor Air Pollution: Hearing, August 5, 1985, on S. 1198, a Bill to Establish in the Environmental Protection Agency a Program of Research on Indoor Air Quality and for Other Purposes.* 99th Cong., 1st sess. Washington, DC: GPO, 1985. 68 pp.

Radon

Alter, H. W., and R. L. Fleischer. "Passive Integrating Radon Monitor for Environmental Monitoring." *Health Physics* 40 (May 1981): 693–702.

Brambley, M. R., and M. Gorfien. "Radon and Lung Cancer: Incremental Risks Associated with Residential Weatherization." *Energy* 11 (June 1986): 589–605.

Castren, O., and others. "Studies of High Indoor Radon Areas in Finland." *Science of the Total Environment* 45 (October 1985): 311–318.

Chittaporn, P., M. Eisenbud, and N. H. Harley. "A Continuous Monitor for the Measurement of Environmental Radon." *Health Physics* 41 (August 1981): 405–410.

Cohen, Bernard L. "A National Survey of ^{222}Rn in U.S. Homes and Correlating Factors." *Health Physics* 51 (August 1986): 175–183.

Cohen, Bernard L., and Ernest S. Cohen. "Theory and Practice of Radon Monitoring with Charcoal Adsorption." *Health Physics* 45 (August 1983): 501–508.

Damkjaer, A., and U. Korsbech. "Measurement of the Emanation of Radon-222 from Danish Soils." *Science of the Total Environment* 45 (October 1985): 343–350.

Fleischer, R., and L. Turner. "Indoor Radon Measurement in the New York Capital District." *Health Physics* 46 (May 1984): 999–1011.

George, Andreas C. "Passive Integrated Measurement of Indoor Radon Using Activated Carbon." *Health Physics* 46 (April 1984): 867–872.

Hagberg, N. "Some Tests on Measuring Methods for Indoor Radon Using Activated Charcoal." *Science of the Total Environment* 45 (October 1985): 417–423.

Harley, Naomi, and Bernard S. Pasternack. "Environmental Radon Daughter Alpha Dose Factors in a Five-Lobed Human Lung." *Health Physics* 42 (June 1982): 789-799.

_____ . "A Model for Predicting Lung Cancer Risks Induced by Environmental Levels of Radon Daughters." *Health Physics* 40 (March 1981): 307-316.

Jacobi, W., and H. G. Paretzke. "Risk Assessment for Indo~ ~ Radon Daughters." *Scien~~ ~~ ~~ ~~~~ ~~~.
~~~ ~~~

~~~~~~en, N., and J. P. McLaughlin. "The Reduction of Indoor Air Concentrations of Radon Daughters Without the Use of Ventilation." *Science of the Total Environment* 45 (October 1985): 485-492.

Myrick, T. E., B. A. Berven, and F. F. Haywood. "Determination of Concentrations of Selected Radionuclides in Surface Soil in the U.S." *Health Physics* 45 (September 1983): 631-642.

Nazaroff, W. W. "Optimizing the Total-Alpha Three-Count Technique for Measuring Concentrations of Radon Progeny in Residences." *Health Physics* 46 (February 1984): 395-405.

Nazaroff, W. W., and others. "Radon Transport into a Detached One-Story House with a Basement." *Atmospheric Environment* 19:1 (1985): 31-46; Discussion 20:5 (1986): 1065-1067.

Nero, A. V., and W. W. Nozaroff. "Characterizing the Source of Radon Indoors." *Radiation Protection Dosimetry* 7 (1984): 23-39.

Nyberg, P., and D. Bernhardt. "Measurement of Time-Integrated Radon Concentrations in Residences." *Health Physics* 45 (August 1983): 539-543.

Piffijn, A., and others. "Results of a Preliminary Survey of Radon in Belgium." *Science of the Total Environment* 45 (October 1985): 335-342.

Proposed Standard for Radon-222 Emissions from Licensed Uranium Mill Tailings, Draft Economic Analysis. Prepared by Jack Faucett Associates. Washington, DC: U.S. Environmental Protection Agency, Office of Radiation Programs, 1986. 199 pp.

Put, L. W., R. J. DeMeijer, and B. Hogeweg. "Survey of Radon Concentrations in Dutch Dwellings." *Science of the Total Environment* 45 (October 1985): 441-448.

"Radon: Pinpointing a Mystery." *EPA (Environmental Protection Agency) Journal* 12 (August 1986): 2-15.

Radon Reduction Techniques for Detached Houses: Technical Guidance. Research Triangle Park, NC: Air and Energy Engineering Research Laboratory, Office of Environmental Engineering and Technology,

Office of Research and Development, U.S. Environmental Protection Agency, 1986. 50 pp.

Reineking, A., K. H. Becker, and J. Portstendörfer." Measurements of the Unattached Fractions of Radon Daughters in Houses." *Science of the Total Environment* 45 (October 1985): 261–270.

Ronca, Battista, and others. *Interim Indoor Radon and Radon Decay Product Measurements Protocols.* Washington, DC: U.S. Environmental Protection Agency, Office of Radiation Programs, 1986. 1 vol. (various paging).

Rudnick, S. N., and E. F. Maher. "Surface Deposition of ^{222}Rn Decay Products with and without Enhanced Air Motion." *Health Physics* 51 (September 1986): 283–293.

Schery, S. D., D. H. Gaeddett, and M. H. Wilkening. "Factors Affecting Exhalation of Radon from Gravelly Sandy Loam." *Journal of Geophysics Research* 89 (1984): 7299–7309.

Schmier, H., and A. Wicke. "Results from a Survey of Indoor Radon Exposures in the Federal Republic of Germany." *Science of the Total Environment* 45 (October 1985): 307–310.

Steinhäusler, F., and others. "Radiation Exposure of the Respiratory Tract and Associated Carcinogenic Risk Due to Inhaled Radon Daughters." *Health Physics* 45 (August 1983): 331–337.

Stranden, E. "Thoron and Radon Daughters in Different Atmospheres." *Health Physics* 38 (May 1980): 777–785.

U.S. Congress, House, Committee on Energy and Commerce, Subcommittee on Transportation, Tourism, and Hazardous Materials. *Radon Pollution Control Act of 1987: Hearing, April 23, 1987.* 100th Cong., 1st sess. Washington, DC: GPO, 1987. 119 pp.

U.S. Congress, House, Committee on Science and Technology. *Radon and Indoor Air Pollution: Hearing, October 10, 1985, Before the Subcommittee on Natural Resources, Agriculture Research, and Environment.* 99th Cong., 1st sess. Washington, DC: GPO, 1986. 291 pp.

U.S. Congress, House, Committee on Science and Technology, Subcommittee on Natural Resources, Agriculture Research, and Environment. *Radon and Indoor Air Pollution: Hearing, October 10, 1985.* 99th Cong., 1st sess. Washington, DC: GPO, 1986. 291 pp.

————. *Residential Radon Contamination and Indoor Quality Research Needs: Hearing, September 17, 1986.* 99th Cong., 2d sess. Washington, DC: GPO, 1987. 333 pp.

U.S. Congress, Senate, Committee on Environment and Public Works. *Radon Gas Issues: Joint Hearings, April 2, 1987, on S. 743 and S. 744,*

Before the Subcommittee on Environmental Protection and Superfund and Environmental Oversight. 100th Cong., 1st sess. Washington, DC: GPO, 1987. 124 pp.

U.S. Congress, Senate, Committee on Environment and Public Works, Subcommittee on Environmental Protection. *Health Effects of Indoor Air Pollution: Hearing, April 24, 1987.* 100th Cong., 1st sess. Washington, DC: GPO, 1987. 124 pp.

~~Van der Laan...~~ *Science of the Total Environment* 45 (October 1985): 143–150.

Vanmarcke, H., A. Janssens, and F. Raes. "The Equilibrium of Attached and Unattached Radon Daughters in the Domestic Environment." *Science of the Total Environment* 45 (October 1985): 251–260.

Vohra, K. G. "An Experimental Study of the Role of Radon and Its Daughter Products in the Conversion of Sulphur Dioxide into Aerosol Particles in the Atmosphere." *Atmospheric Environment* 18:8 (1984): 1653–1665.

Wilkening, Marvin, and Andreas Wicke. "Seasonal Variation of Indoor Rn at a Location in the Southwestern United States." *Health Physics* 51 (October 1986): 427–436.

Wilkinson, P., and P. J. Dimbylon. "Radon Diffusion Modelling." *Science of the Total Environment* 45 (October 1985): 227–232.

Wilkinson, P., and B. J. Saunders. "Theoretical Aspects of the Design of a Passive Radon Dosemeter." *Science of the Total Environment* 45 (October 1985): 433–440.

Wilkniss, P. E., and R. E. Larson. "Atmospheric Radon Measurements in the Arctic: Fronts, Seasonal Observations, and Transport of Continental Air to Polar Regions." *Journal of the Atmospheric Sciences* 41 (August 1, 1984): 2347–2358.

Smoking

Burns, T. R., and others. "Smoke Inhalation: An Ultrastructural Study of Reaction to Injury in the Human Alveolar Wall." *Environmental Research* 41 (December 1986): 447–457.

Cain, W. S., and others. "Environmental Tobacco Smoke: Sensory Reactions of Occupants." *Atmospheric Environment* 21:2 (1987) 347–353.

Churg, A., and B. Wiggs, "Types, Numbers, Sizes, and Distribution of Mineral Particles in the Lungs of Urban Male Cigarette Smokers." *Environmental Research* 42 (February 1987): 121–129.

Hammond, S. K., and others. "Collection and Analysis of Nicotine as a Marker for Environmental Tobacco Smoke." *Atmospheric Environment* 21:2 (1987): 457–462.

Muramatsu, M., and others. "Estimation of Personal Exposure to Tobacco Smoke with a Newly Developed Nicotine Personal Monitor." *Environmental Research* 35 (October 1984): 218–227.

Offermann, F. J., and others. "Control of Respirable Particles in Indoor Air with Portable Air Cleaners." *Atmospheric Environment* 19:11 (1985): 1761–1771.

Schein, David D. "Should Employers Restrict Smoking in the Workplace?" *Labor Law Journal* 38 (March 1987): 173–178.

U.S. Congress, House, Committee on Energy and Commerce, Subcommittee on Health and the Environment. *Designation of Smoking Areas in Federal Buildings: Hearings, June 12 and 27, 1986, on H.R. 4488 and H.R. 4546, Bills to Restrict Smoking in Designated Areas in All Buildings or Building Sections Occupied by the U.S. Government.* 99th Cong., 2d sess. Washington, DC: GPO, 1987. 874 pp.

Laws and Regulations

Virginia, Air Pollution Control Board. *Regulations for the Control and Abatement of Air Pollution.* Richmond, VA: 1981. 229 pp.

Clean Air Act

"The Acid Rain Controversy: Pro & Con." *Congressional Digest* 64 (February 1985): 33–64.

"Air Quality in the Southwest." Seven Articles. *Southwestern Review of Management and Economics* 1 (Summer 1982): 1–163.

Buche, Craig Michael. "State Implementation Plans Under the Clean Air Act: Continued Enforceability as Federal Law, After State Court Invalidation on State Grounds." *Valparaiso University Law Review* 19 (Summer 1985): 877–898.

Butler, Chad. "New Source Netting in Nonattainment Areas Under the Clean Air Act." *Ecology Law Quarterly* 11:3 (1984): 343–372.

California, Laws, Statutes, etc. *California Air Pollution Control Laws, 1980.* Sacramento, CA: California Air Resources Board, 1980. 109 pp.

_____ . *California Air Pollution Control Laws, 1981.* Sacramento, CA: California Air Resources Board, 1981. 114 pp.

Calvo y Gonzalez, Jorge A. del. "Markets in Air: Problems and Prospects of Controlled Trading." *Harvard Environmental Law Review* 5:2 (1981): 377–430.

Cannon, J. A. "The Regulation of Toxic Air Pollutants." *Journal of the Air Pollution Control Association* 36 (March 1986): 562–573; Discussion 36 (September 1986): 986–996.

"Clean Air Guide." *State Government News* 24 (April 1981): 4-24.

"Controversy over the Clean Air Act: Pro and Con." *Congressional Digest* 61 (January 1982): 3-32.

Cook, Kenneth F., "Pollution Control and Intergovernmental Relations: Is the Federal 'Stick' Necesary?" In Judd, Dennis R., ed., *Public Policy Across States and Communities,* pp. 91-107. Greenwich, CT: JAI Press, 1985.

D.lD   *E.n.o.r..mental Law Review* 5:1 (1981): 184-203.

Durant, Robert F., and others. "When Government Regulates Itself: The EPA/TVA [Environmental Protection Agency/Tennessee Valley Authority] Air Pollution Control Experience." *Public Administration Review* 43 (May-June 1983): 209-219.

Fisk, David P. "Will Acid Rain on a Modest Proposal?" *Glendale Law Review* 6:2 (1984): 108-124.

Floy, Kent R. "Prevention of Significant Deterioration of Air Quality—The Regulations After Alabama Power." *Boston College Environmental Affairs Law Review* 9:1 (1980-1981): 13-61.

Gallogly, Margaret R. "Acid Precipitation: Can the Clean Air Act Handle It?" *Boston College Environmental Affairs Law Review* 9:3 (1981): 687-744.

Glusco, Sheila A. "Proposed Clean Air Act Amendments: The United States Response to Acid Rain." *George Washington Journal of International Law and Economics* 17:1 (1982): 137-166.

Hahn, Robert W., and Roger G. Noll. "Barriers to Implementing Tradable Air Pollution Permits: Problems of Regulatory Interactions [California]." *Yale Journal on Regulation* 1:1 (1983): 63-91.

Kalikow, B. N. "Environmental Risk: Power to the People [Tacoma, WA]." *Technology Review* 87 (October 1984): 54-61.

Landau, Jack L. "Economic Dream or Environmental Nightmare? The Legality of the 'Bubble Concept' in Air and Water Pollution Control." *Boston College Environmental Affairs Law Review* 8:4 (1980): 741-781.

Lind, Douglas. "Umbrella Equities: Use of the Federal Common Law of Nuisance to Catch the Fall of Acid Rain." *Urban Law Annual* 21 (1981): 143-178.

Milstone, Nancy Hughes. "A Common Law Solution to the Acid Rain Problem." *Valparaiso University Law Review* 20 (Winter 1986): 277-297.

Navarro, Peter. "The 1977 Clean Air Act Amendments: Energy, Environmental, Economic, and Distributional Impacts of Title I." *Public Policy* 29 (Spring 1981): 121-146.

Pashigian, B. Peter. "Environmental Regulation: Whose Self-Interests Are Being Protected?" *Economic Inquiry* 23 (October 1985): 551-584.

Reed, Phillip D. "The Trial of Hazardous Air Pollution Regulation." *Environmental Law Reporter* 16:3 (1986): 10066-10072.

_____. "When Is an Area That Is in Attainment Not an Attainment Area?" *Environmental Law Reporter* 16 (February 1986): 1041-1048.

Russell, M., and J. Greenwald. "Environmental Policies Affecting the Development of Newer Coal Technologies." *Energy* 11 (November-December 1986): 1301-1314.

Schoenbrod, David. "Goals Statutes or Rules Statutes: The Case of the Clean Air Act." *UCLA Law Review* 30 (April 1983): 740-828.

Stewart, Thomas R., and others. "Citizen Participation and Judgment in Policy Analysis: A Case Study of Urban Air Quality Policy [Colorado]." *Policy Sciences* (Amsterdam) 17 (May 1984): 67-87.

Thornton, Charles A., and Robert L. Koepke. "Federal Legislation, Clean Air, and Local Industry." *Geographical Review* 71 (July 1981): 324-339.

Trisko, Eugene M. "Clean Air and Electric Power Developments." *Public Utilities Fortnightly* 109 (January 21, 1982): 24-31.

U.S. Congress, House, Committee on Energy and Commerce. *Acid Rain in the West: Hearing, June 28, 1985, Before the Subcommittee on Health and the Environment.* 99th Cong., 1st sess. Washington, DC: GPO, 1986. 315 pp.

U.S. Congress, House, Committee on Energy and Commerce, Subcommittee on Energy Conservation and Power. *Fuel Use Act Amendments: Hearing, June 2, 1981, on H.R. 1765, a Bill to Reform the Powerplant and Industrial Fuel Use Act of 1978 to Encourage a Reduction of Air Pollution and Oil Consumption by Existing Electric Powerplants.* 97th Cong., 1st sess. Washington, DC: GPO, 1981. 296 pp.

U.S. Congress, House, Committee on Energy and Commerce, Subcommittee on Health and the Environment. *Acid Deposition Control Act of 1986: Hearings, April 29-May 7, 1986, on H.R. 4567, a Bill to Amend the Clean Air Act to Reduce Acid Deposition, and for Other Purposes.* 99th Cong., 2d sess. Washington, DC: GPO, 1986. 3 pts.

_____. *Acid Precipitation: Hearings: pts. 1-2, October 1-20, 1981, on Effects and Solutions to Combat Acid Precipitation.* 97th Cong., 1st sess. Washington, DC: GPO, 1982. 2 pts.

_____. *Acid Rain Control: Hearings, pts. 1-2, December 1, 1983-March 5, 1984, on H.R. 3400, a Bill to Amend the Clean Air Act to Control Certain Sources of Sulfur Dioxides and Nitrogen Oxides to Reduce Acid Deposition and for Other Purposes.* 98th Cong., 1st and 2d sess. Washington, DC: GPO, 1984. 2 pts.

_____ . *Acid Rain in Europe: A Report, March 1985, on the Acid Rain Fact-Finding Excursion.* 99th Cong., 1st sess. Washington, DC: GPO, 1985. 97 pp.

_____ . *Clean Air Act: Hearings, pts. 1–3, October 22, 1981–Februry 23, 1982.* 97th Cong., 1st and 2d sess. Washington, DC: GPO, 1982. 3 pts.

_____ . *Mobile Source Provisions: Hearings, September 21, 1981–January 21, 1982, on H.R. 4400 and H.R. 2310* [Bill . . . . and to . . . Consumer Freedom of Choice to Select Parts and Service of the Consumer's Own Choosing and for Other Purposes.* 97th Cong., 1st and 2d sess. Washington, DC: GPO, 1982. 1519 pp.

U.S. Congress, House, Committee on Energy and Commerce, Subcommittee on Oversight and Investigations. *EPA [Environmental Protection Agency]'s Pollution Control Program: Hearing, November 7, 1983.* 98th Cong., 1st sess. Washington, DC: GPO, 1984. 904 pp.

U.S. Congress, House, Committee on Foreign Affairs. *United States–Canadian Relations and Acid Rain: Hearing, May 20, 1981, Before the Subcommittees on Human Rights and International Organizations and on Inter-American Affairs.* Washington, DC: GPO, 1981. 121 pp.

U.S. Congress, House, Committee on Interior and Insular Affairs, Subcommittee on Mining, Forest Management, and Bonneville Power Administration. *Effects of Air Pollution and Acid Rain on Forest Decline: Oversight Hearing, June 7, 1984.* 98th Cong., 2d sess. Washington, DC: GPO, 1984. 230 pp.

U.S. Congress, House, Committee on Interstate and Foreign Commerce, Subcommittee on Health and the Environment. *Clean Air Act Oversight—1980: Hearing, June 16, 1980, on Greater Flexibility in Meeting Pollution Standards and Aspects of the Prevention of Significant Deterioration.* 96th Cong., 2d sess. Washington, DC: GPO, 1980. 218 pp.

U.S. Congress, House, Committee on Interstate and Foreign Commerce, Subcommittee on Oversight and Investigations. *Acid Rain: Hearings, February 26–27, 1980.* 96th Cong., 2d sess. Washington, DC: GPO, 1980. 784 pp.

U.S. Congress, House, Committee on Science and Technology. *Acid Rain—Implications for Fossil R&D: Hearings, September 13 and 20, 1983, Before the Subcommittee on Energy Development and Applications and the Subcommittee on Natural Resources, Agriculture Research, and Environment.* 98th Cong., 1st sess. Washington, DC: GPO, 1984. 1287 pp.

_____ . *Clean Air Act: Hearings, May 19–28, 1981, Before the Subcommittee on Energy Development and Applications and the Subcommittee on Natural Resources, Agriculture Research, and Environment.* 97th Cong., 1st sess. Washington, DC: GPO, 1982. 354 pp.

U.S. Congress, House, Committee on Science and Technology, Subcommittee on Natural Resources, Agriculture Research, and Environment. *Acid Rain: Hearings, September 18–December 9, 1981.* 97th Cong., 1st sess. Washington, DC: GPO, 1982. 678 pp.

————. *Mitigating Acid Rain with Technology: Avoiding the Scrubbing-Switching Dilemma, Report June 1983.* Larry B. Parker and Robert E. Trumbule. 98th Cong., 1st sess. Washington, DC: GPO, 1983. 813 pp.

U.S. Congress, House, Select Committee on Aging, Subcommittee on Human Services. *Alzheimer's Disease: Is There an Acid Rain Connection? Hearing, August 8, 1983.* 98th Cong., 1st sess. Washington, DC: GPO, 1983. 89 pp.

U.S. Congress, Office of Technology Assessment. *Acid Rain and Transported Air Pollutants: Implications for Public Policy.* New York: UNIPUB, 1985. 323 pp.

U.S. Congress. Senate, *Economic Impact of Acid Rain: Hearing, September 23, 1980, Before the Select Committee on Small Business and the Committee on Environment and Public Works.* 96th Cong., 2d sess. Washington, DC: GPO, 1980. 224 pp.

U.S. Congress, Senate, Committee on Energy and Natural Resources. *Acid Precipitation and the Use of Fossil Fuels: Hearing, August 19, 1982, to Review the Issue of Acid Precipitation and Fossil Fuel Use in Our Economy.* 97th Cong., 2d sess. Washington, DC: GPO, 1982. 1542 pp.

————. *Clean Cool Technology Development and Strategies for Acid Rain Control: Hearings, June 9 and 10, 1986.* 99th Cong., 2d sess. Washington, DC: GPO, 1987. 935 pp.

————. *Effects of Acid Rain: Hearing, pt. 1, May 28, 1980, on the Phenomenon of Acid Rain and Its Implications for a National Energy Policy.* 96th Cong., 2d sess. Washington, DC: GPO, 1980. 752 pp.

————. *Implementation of the Acid Precipitation Act of 1980: Hearing, April 30, 1984.* 98th Cong., 2d sess. Washington, DC: GPO, 1984. 1806 pp.

————. *Summary of Oversight Hearings on Implementation of the Acid Precipitation Act of 1980.* Prepared by Larry B. Parker and others. 98th Cong., 2d sess. Washington, DC: GPO, 1984. 70 pp.

U.S. Congress, Senate, Committee on Energy and Natural Resources, Subcommittee on Energy Conservation and Supply. *Effects of Acid Rain: Hearing, pt. 2, June 21, 1980, on the Nature, Source, and Effects of Acid Rain and How the Potential Growth in Emissions from Coal Burning Powerplants Will Affect Acid Rain Problems.* 96th Cong., 2d sess. Washington, DC: GPO, 1981. 121 pp.

U.S. Congress, Senate, Committee on Environment and Public Works. *Acid Rain: Hearing, October 29, 1981, on S. 1706 and Other Bills.* 97th Cong., 1st sess. Washington, DC: GPO, 1982. 787 pp.

———. *Acid Rain in the West: Hearing, August 12, 1985.* 99th Cong., 1st sess. Washington, DC: GPO, 1986. 285 pp.

———. *Acid Rain, 1984: Hearings, February 2–10, 1984.* 98th Cong., 2d sess. Washington, DC: GPO, 1984. 735 pp.

———. *Acid Rain—A Technical Inquiry: Hearings, May 25 and 27, 1982.* 97th Cong., 2d sess. Washington, DC: GPO, 1982. 848 pp.

———. *Clean Air Act Amendm~~~~~~~~~~~~~~~~

~~~~~~~~~~~~~~~~~~~~~~~~~~~~~g., 1st sess. Wash-

~~~~~~~~~~, 1984. 2 pts.

———. *The Clean Air Act in the Courts: A Report.* Edited by Robert Meltz. 97th Cong., 1st sess. Washington, DC: GPO, 1981. 164 pp.

———. *Clean Air Act Oversight: Hearings: pts. 1–6, April 8–July 9, 1981.* 97th Cong., 1st sess. Washington, DC: GPO, 1981. 6 pts.

———. *Federal-State Relations in Transition: Implications for Environmental Policy, a Report.* Claudia Copeland. 97th Cong., 2d sess. Washington, DC: GPO, 1982. 99 pp.

———. *The New Clean Air Act: Hearings, September 25–October 2, 1986, on S. 2203, a Bill to Establish a Program to Reduce Acid Deposition, and for Other Purposes.* 99th Cong., 2d sess. Washington, DC: GPO, 1986. 671 pp.

U.S. Laws, Statutes, etc. *The Clean Air Act as Amended, August 1977 and July 1980.* 96th Cong., 2d sess. Washington, DC: Committee on Environment and Public Works, U.S. Senate, 1980. 185 pp.

———. *The Clean Air Act as Amended Through July 1981.* 97th Cong., 1st sess. Washington, DC: Committee on Environment and Public Works, U.S. Senate, 1981. 190 pp.

U.S., National Commission on Air Quality. *To Breathe Clean Air: Report.* Washington, DC: Superintendent of Documents, 1981. 356 pp.

Yandle, Bruce. "Sulfur Dioxide: State Versus Federal Control." *Journal of Energy and Development* 10 (Autumn 1984): 63–72.

International Aspects

Brockmann, James C. "Acid Rain: Corroding United States–Canadian Relations." *Journal of Energy Law and Policy* 6:2 (1985): 357–390.

Brown, Susan. "International–United States Air Pollution Control and the Acid Rain Phenomenon." *Natural Resources Journal* 21 (July 1981): 631–645.

Caplan, Bennett A. "The Applicability of Clean Air Act Section 115 to Canada's Transboundary Acid Precipitation Problem." *Boston College Environmental Affairs Law Review* 11 (April 1984): 539–607.

Dente, Bruno, and Rudy Lewanski. "Administrative Networks and Implementation Effectiveness: Industrial Air Pollution Control Policy in Italy." *Policy Studies Journal* 11 (September 1982): 116–129.

Handl, Günther. "National Uses of Transboundary Air Resources: The International Entitlement Issue Reconsidered." *Natural Resources Journal* 26 (Summer 1986): 405–467.

Hart, David. "Acid Deposition: the European Situation." *Forum for Applied Research and Public Policy* 2 (Fall 1987): 58–67.

Johnston, Douglas M., and Peter Finkle. "Acid Precipitation in North America: The Case for Transboundary Cooperation." *Vanderbilt Journal of Transnational Law* 14 (Fall 1981): 787–843.

Khordokovskaia, S. "The Effectiveness of Protection of the Atmosphere [Soviet Union]." *Problems of Economics* 27 (March 1985): 59–74.

Knoepfel, Peter, and Helmut Weidner. "Implementing Air Quality Control Programs in Europe: Some Results of a Comparative Study." *Policy Studies Journal* 11 (September 1982): 103–115.

Lyle, John B. "International Liability and Primary Rules of Obligation: An Application to Acid Rain in the United States and Canada." *Georgia Journal of International and Comparative Law* 13 (Winter 1983): 111–134.

Mingst, Karen A. "Evaluating Public and Private Approaches to International Solutions to Acid Rain Pollution." *Natural Resources Journal* 22 (January 1982): 5–20.

Mumme, Steven P. "The Cananea Copper Controversy: Lessons for Environmental Diplomacy." *Inter-American Economic Affairs* 38 (Summer 1984): 3–22.

Ray, Dixie Lee. "The Great Acid Rain Debate: No One in Washington (or Ottawa) Knows What He's Talking About." *American Spectator* 20 (January 1987): 21–25.

Rosencranz, Armin. "The International Law and Politics of Acid Rain." *Denver Journal of International Law and Policy* 10 (Spring 1981): 511–521.

Scott, Anthony. "The Canadian-American Problems of Acid Rain." *Natural Resources Journal* 26 (Spring 1986): 337–358.

Stanfield, Rochelle L. "Antacid Remedy: With Another Debate on Acid Rain Coming Up in Congress, West Germany Offers Lessons in Pollution Control; But It Took Trouble in the Black Forest to Get Action." *National Journal* 19 (June 27, 1987): 1655–1659.

Sullivan, John L. "Beyond the Bargaining Table: Canada's Use of Section 115 of the United States Clean Air Act to Prevent Acid Rain." *Cornell International Law Journal* 16 (Winter 1983): 193–227.

Thompson, Roger. "Acid Rain: Canada's Push for U.S. Action." *Editorial Research Reports* (March 7, 1986): 167–184.

United Nations, Economic Commission for Europe. *Transboundary Air Pollution: Effects and Control.* Air Pollution Studies no. 3. New York: UN, 1986. 77 pp.

U.S. Congress, Senate, Committee on Environment and Public Works. *Air Quality Management ~~~~~~~~~ pp.*

Wetstone, Gregory, and Armin Rosencranz. "Transboundary Air Pollution: The Search for an International Response." *Harvard Environmental Law Review* 8:1 (1984): 89–138.

Wilcher, Marshall E. "The Acid Rain Debate in North America: 'Where You Stand Depends on Where You Sit.'" *Environmentalist* 6 (Winter 1986): 289–298.

Selected Journal Titles

The journals listed below publish articles on many aspects of air pollution. Because the problems of air pollution have become important only in recent years, new journals are continually appearing. For new journals and additional information please consult:

Ulrich's International Periodicals Directory
1987–1988. 26th edition.
New York: R. R. Bowker Company, 1987. 2 vols.

Information on the journal listed is arranged in the following manner:

Sample Entry
Journal Title
1. Editor
2. Year first published
3. Frequency of publication
4. Code
5. Special features
6. Address of publisher

Acid Precipitation Digest
1. Mary T. Pratt
2. 1983

3. Monthly
4. ISSN 0740-2252
5. Book reviews
6. Center for Environmental Information, Inc.
   33 S. Washington Street
   Rochester, NY 14608-2046

Air & Water Pollution Control

1. Eileen Z. Joseph
2. 1986
3. Biweekly
4. ISSN 0890-0396
5. _____
6. Bureau of National Affairs, Inc.
   1231 25th Street, N.W.
   Washington, DC 20037

Air Pollution Control

1. Eileen Z. Joseph
2. 1980
3. Biweekly
4. ISSN 0196-7150
5. Index
6. Bureau of National Affairs, Inc.
   1231 25th Street, N.W.
   Washington, DC 20037

Air Pollution Control Association Journal

1. Harold M. Englund
2. 1951
3. Monthly
4. ISSN 0002-2470
5. Directory of government air pollution agencies
6. Air Pollution Control Association
   Three Gateway Center
   Four West
   Pittsburgh, PA 15222

American Industrial Hygiene Association Journal

1. Paul D. Halley
2. 1940
3. Monthly
4. ISSN 0002-8894
5. Adv., bibl., charts, illus., index, cum. index 1940–1957
6. American Industrial Hygiene Association
   475 Wolfledges Parkway
   Akron, OH 44311

Atmospheric Environment

1. James P. Lodge, Jr.
2. 1967
3. Monthly
4. ISSN 0004-6981
5. Adv., bk. rev., charts, illus.
6. Pergamon Press, Inc.
   Journals Division

   Elmsford, NY 10523

Boston College Environmental Affairs Law Review

1. Editorial Board
2. 1971
3. Quarterly
4. ISSN 0190-7034
5. Adv., bk. rev., bibl., charts, index
6. School of Law
   Boston College
   855 Centre Street
   Newton Centre, MA 02159

Bulletin of the American Meteorological Society

1. Kenneth C. Spengler
2. 1920
3. Monthly
4. ISSN 0003-0007
5. Adv., bk. rev., bibl., charts, illus., index
6. American Meteorological Society
   45 Beacon Street
   Boston, MA 02108

Ecology Law Quarterly

1. Beverly Alexander
2. 1971
3. Quarterly
4. ISSN 0046-1121
5. Adv., bk. rev., bibl., index
6. University of California Press
   Journals Division
   2120 Berkeley Way
   Berkeley, CA 94720

Energy Journal

1. Helmut Frank
2. 1980
3. Quarterly

    4. ISSN 0195-6574
    5. Adv., bk. rev., charts, index
    6. International Association of Energy Economists
       131 Clarendon Street
       Boston, MA 02116

Energy Policy: The Political, Economics, Planning, and Social Aspects of Energy

    1. _____
    2. 1973
    3. Bimonthly
    4. UK ISSN 0301-4215
    5. Adv., bk. rev., charts, illus., stat., index
    6. Butterworth Scientific, Ltd.
       P.O. Box 63
       Westbury House
       Bury Street
       Guildford, Surrey GU2 5BH
       England

Environmental Action

    1. Editorial Board
    2. 1970
    3. Bimonthly
    4. ISSN 0013-922X
    5. Adv., bk. rev., film rev., illus., index
    6. Environmental Action, Inc.
       1525 New Hampshire Ave., N.W.
       Washington, DC 20036

Environmental Forum

    1. Morris A. Ward
    2. _____
    3. Monthly
    4. ISSN 0731-5732
    5. Index
    6. Environmental Law Institute
       1616 P Street, N.W.
       Suite 200
       Washington, DC 20036

Environmental Impact Assessment Review

    1. Lawrence E. Susskind
    2. 1980–1983, resumed 1986
    3. Quarterly
    4. ISSN 0195-9255
    5. Index

6. Elsevier Science Publishing Company, Inc.
   52 Vanderbilt Avenue
   New York, NY 10017

Environmentalist

1. John F. Potter
2. 1980
3. Quarterly

6. Science and Technology Letters
   12 Clarence Road
   Kew, Surrey TW9 3NL
   England

Environmental Law Reporter

1. Phillip D. Reed
2. 1971
3. Monthly
4. ISSN 0046-2284
5. Bk. rev., index
6. Environmental Law Institute
   1616 P Street, N.W.
   Suite 200
   Washington, DC 20036

Environmental Management

1. David Alexander
2. 1977
3. Bimonthly
4. ISSN 0364-152X
5. Bk. rev., charts
6. Springer-Verlag
   175 Fifth Avenue
   New York, NY 10010

Environmental Monitoring and Assessment

1. G. Bruce Wiersma
2. 1981
3. Six times per year
4. NE ISSN 0167-6369
5. Adv., bk. rev., illus.
6. D. Reidel Publishing Company
   Box 17
   3300 AA Dordrecht
   Netherlands

Environmental Policy and Law

1. Wolfgang E. Burhenne
2. 1975
3. Six times per year
4. NE ISSN 0378-777X
5. Adv., bk. rev., bibl., charts, illus., index
6. Elsevier Science Publishers, B.V.
   Box 211
   1100 AE Amsterdam
   Netherlands

Environmental Research

1. I.J. Selikoff
2. 1967
3. Bimonthly
4. ISSN 0013-9351
5. Adv., bk. rev., illus., index
6. Academic Press, Inc.
   Journal Division
   1250 Sixth Avenue
   San Diego, CA 92101

Environmental Science & Technology

1. Russell F. Christman
2. 1967
3. Monthly
4. ISSN 0013-936X
5. Adv., bk. rev., abstr., bibl., charts, illus. stat., index
6. American Chemical Society
   1155 16th Street, N.W.
   Washington, DC 20036

EPA (Environmental Protection Agency) Journal

1. John Heritage
2. 1975
3. Quarterly
4. _____
5. Adv., index
6. Environmental Protection Agency
   Office of Public Affairs
   Waterside Mall
   401 M Street, S.W.
   Washington, DC 20460

Harvard Environmental Law Review

1. Editorial Board
2. 1976

3. Two times per year
4. ISSN 0147-8257
5. Index
6. Harvard University Law School
   Publication Center
   Cambridge, MA 02138

Journal of Climate and Applied Meteorology

1. Bernard Sil

3. Monthly
4. ISSN 0733-3021
5. Abst., bibl., charts, illus., index
6. American Meteorological Society
   45 Beacon Street
   Boston, MA 02108

Journal of Energy and Development

1. Ragaeiel Mallakh
2. 1975
3. Twice annually
4. ISSN 0361-4476
5. Adv., bk. rev., charts, stat.
6. International Research Center for Energy and Economic
      Development
   216 Economics Building, Box 263
   University of Colorado
   Boulder, CO 80309-0263

Journal of Energy Law and Policy

1. Editorial Board
2. 1980
3. Twice annually
4. ISSN 0275-9926
5. Bk. rev., index
6. College of Law
   University of Utah
   Salt Lake City, UT 84112

Journal of Environmental Economics and Management

1. Ronald G. Cummings
2. 1975
3. Quarterly
4. ISSN 0095-0696
5. Index
6. Academic Press, Inc.
   Journal Division

1250 Sixth Avenue
San Diego, CA 92101

Journal of Environmental Engineering

1. Editorial Board
2. 1956
3. Bimonthly
4. ISSN 0733-9372
5. Index
6. American Society of Civil Engineers
   345 East 47th Street
   New York, NY 10017

Journal of the Atmospheric Sciences

1. J. M. Fritch and P. J. Webster
2. 1944
3. Twice monthly
4. ISSN 0022-4928
5. Charts, illus., index, stat.
6. American Meterorological Society
   45 Beacon Street
   Boston, MA 02108

Monthly Weather Review

1. Roger Pielke
2. 1872
3. Monthly
4. ISSN 0027-0644
5. Charts, illus., stat., index
6. American Meteorological Society
   45 Beacon Street
   Boston, MA 02108

Natural Resources Journal

1. Albert E. Utton
2. 1961
3. Quarterly
4. ISSN 0028-0739
5. Adv., bk. rev., charts, index, cum. index every ten years
6. University of New Mexico
   School of Law
   1117 Stanford NE
   Albuquerque, NM 87131

Nuclear Technology

1. Roy G. Post
2. 1965
3. Monthly

4. ISSN 0029-5450
5. Bk. rev., charts, illus., stat., index
6. American Nuclear Society
   555 N. Kensington Avenue
   La Grange Park, IL 60525

Pollution Engineering

1. Richard Young

4. _____
5. _____
6. Pudvan Publishing Company
   1935 Shermer Road
   Northbrook, IL 60062

Population and Environment

1. Burton Mindick and Ralph Taylor
2. 1978
3. Quarterly
4. ISSN 0199-0039
5. Adv., bk. rev., charts, index
6. Human Sciences Press, Inc.
   72 Fifth Avenue
   New York, NY 10011

Public Utilities Fortnightly

1. Lucien Smartt
2. 1929
3. Fortnightly
4. ISSN 0033-3808
5. Adv., bk. rev., index
6. Public Utilities Reports, Inc.
   Suite 2100 Rosslyn Center Building
   1700 North Moore Street
   Arlington, VA 22209

Radiation Protection Dosimetry

1. T. F. Johns
2. 1981
3. 16 times per year (4 volumes)
4. UK ISSN 0144-8420
5. Adv., bk. rev.
6. Nuclear Technology Publishing
   Box 7
   Ashford, Kent TN25 4NW
   England

Science

1. Daniel Koshland
2. 1880
3. Weekly (4 volumes per year)
4. ISSN 0036-8075
5. Adv., bk. rev., abstr., bibl., illus.
6. American Association for the Advancement of Science
   1333 H Street, N.W.
   Washington, DC 20005

Science of the Total Environment

1. E. I. Hamilton
2. 1972
3. 21 times per year
4. NE ISSN 0048-9697
5. Adv., bk. rev., charts, illus., index
6. Elsevier Science Publishers, B.V.
   Box 211
   1000 AE Amsterdam
   Netherlands

Science, Technology, and Human Values

1. Marcel Chatkowski La Follette
2. 1972
3. Quarterly
4. ISSN 0162-2439
5. Bibl., index
6. John Wiley & Sons, Inc.
   605 Third Avenue
   New York, NY 10158

7

Films, Film Strips, and Videocassettes

THE FILMS AND FILM STRIPS LISTED in this chapter demonstrate that the environmental problems of the atmosphere can be defined and analyzed by using pictures. A graphic presentation conveys the problems more vividly than the written word. For the general public the problems are so subtle that the many forms of air pollution are frequently ignored. As air pollution problems have increased, the number of films has also grown.

The following sources list films, in English, that are produced.

Educational Film/Video Locator of the Consortium of University Film Center and R.R. Bowker, 3rd ed. 2 vols. New York: R. R. Bowker, 1986. 3115 pp.

Films in the Sciences: Reviews and Recommendations. Washington, DC: American Association for the Advancement of Science, 1980. 172 pp.

Index to Environmental Studies (Multimedia). University Park, Los Angeles, CA: National Information Center for Educational Media, University of Southern California, 1977. 1113 pp.

The Video Source Book, 8th ed., Syosset, NY: National Video Clearinghouse, 1986. 2224 pp. and supplements.

General

Pollution: Air, Land, Water, Noise
Academy Films
P.O. Box 3414
Orange, CA 92665
Color, 17 minutes, sound, 16mm., 1971.

This film presents a perspective on human pollution of the environment. The main sources of pollution are shown, and a challenge is given to consider what can be done about the problem. Air pollution in our cities is an obvious problem. However, lake, river, and ocean pollution is also serious.

Pollution Control—Hard Decisions
BNA Communications
9439 Key West Avenue
Rockville, MD 20850
Color, 30 minutes, sound, 16mm., 1971.

Considers the challenges executives face in dealing with problems of the environment. The environmental challenges involve a host of related issues in addition to industrial pollution, such as the destruction of parks, woodlands, seashores, wetlands, and wildlife; impact of growing population, congestion, and noise; and economic security. Executives are required to make difficult and frequently unpopular decisions on the effects of an action on the environment.

Pollution in Paradise
University of Washington, Film Library
Kane Hall DG-10
Seattle, WA 98195
Color, 50 minutes, sound, 16mm., 1963.

Describes and analyzes human pollution of air and water resources in the Pacific Northwest. Political and economic consequences of pollution in Oregon are discussed for several specific cases.

Pollution Is a Matter of Choice
National Broadcasting Company
30 Rockefeller Plaza
New York, NY 10020
Color, 53 minutes, sound, 16mm., 1970.

An NBC White Paper reviews the pollution crisis as a problem in priorities. A choice exists between unlimited technological progress and an abundance of consumer goods on the one hand and the preservation of the environment on the other. The problem is not a simple one. Three communities provide case studies: Should Machiasport, Maine, become an international oilport? Should the proposed Miami jetport save the Everglades? Are the living conditions in Gary, Indiana, too high a price to pay for the nation's steel?

Pollution—It's Up to You

~~30 Rockefeller Plaza~~
New York, NY 10020
Color, 10 minutes, sound, 16mm., 1971.

This film shows that everyone is responsible for the environment and if the ecological balance is to be maintained everyone must make some sacrifice. The progress of environmental degradation can be broken but only if everyone contributes to a better environment.

Pollution of the Upper and Lower Atmosphere
Learning Corporation of America
1350 Avenue of the Americas
New York, NY 10019
Color, 17 minutes, sound, 16mm., 1975.

Shows the effects of the internal combustion engine on the environment and makes the viewer aware of the options with regard to environmental decisions. Emphasizes the implicit dangers in technological expansion. By means of laboratory demonstrations using a small internal combustion engine, the viewer is shown how human pollutants are produced and introduced into the biosphere and the stratosphere. Industrial and natural sources of pollution are given little consideration.

Air Pollution

Air Pollution
Genesys Systems
3788 Fabian Way
Palo Alto, CA 94303
Color, 28 minutes, sound, Beta, VHS, 1/2" reel, 3/4" U-Matic, 1977.

A training film for engineers on the development of responsibility concerning air pollution.

Air Pollution
Sterling Educational Films
241 E. 34th Street
New York, NY 10016
Color, 10 minutes, optical sound, 16mm., 1970.

Explores the causes and results of air pollution, emphasizing that these pollutants know no social or geographical boundaries.

Air Pollution
Visual Education Consultants
Box 52, 2066 Helena Street
Madison, WI 53701
Black and white, filmstrip with script, 1970.

Discusses air pollution, effects of air pollution, and the need for public support in combating pollution.

Air Pollution
Denoyer-Geppert Audio-Visuals
355 Lexington Avenue
New York, NY 10017
Color, filmstrip with tape/script CAP, 1968.

Discusses the origin of air pollution, inversion domes, and effects of pollution on living things and on world temperature.

Air Pollution
Sterling Educational Films
241 E. 34th Street
New York, NY 10016
Color, 9 minutes, sound, 16mm., 1968.

The causes of modern-day pollution from industry, automobiles, and disposal of waste materials are shown. Some possible solutions are given.

Air Pollution
Journal Films
930 Pitner Avenue
Evanston, IL 60202
Color, 10 minutes, sound, 16mm., 1968.

Defines air pollution, shows how it is caused, and presents evidence of the rapidly growing threat it poses to the national health and economy. Discusses the scientific and political complications involved in pollution control and lists possible long-range remedies.

Air Pollution
University Media Extension Technology Planning Center
2378 E. Stadium Boulevard
P.O. Box 1443
Ann Arbor, MI 48106
Color, sound filmstrip-record, n.d.

Defines the nature of the problem of air pollution and identifies the
sources and various types of air pollution
⋯ preventing air pollution.

Air Pollution
Tech Films
222 Arsenal Street
Watertown, MA 02172
Color, 27 minutes, optical sound, 16mm., 1970.

Relates the story of the 1970 clean air race between MIT and Cal Tech
students. Demonstrates that automobile air pollution can be controlled.

Air Pollution: A First Film
BFA Educational Media, Division of Columbia Broadcasting System
470 Park Avenue South
New York, NY 10016
Color, 9 minutes, sound, 16mm., 1971.

Reveals how factories, sawmills, power plants, cars and trucks, road
building, and burning of trash contribute to air pollution. Demonstrates
the harm of air pollution to health, buildings, vegetation, and animals.
Emphasizes that everyone is responsible for air pollution and therefore
must share in correcting the problem.

Air Pollution: A First Film Rev. ed.
BFA Education Media, Division of Columbia Broadcasting System
470 Park Avenue South
New York, NY 10016
Color, 12 minutes, sound, 16mm, ³/₄" U-Matic, ¹/₂" VHS, ¹/₂" Beta, 1984.

Explains the composition and function of the atmosphere showing that
the atmospheric layer is very thin and extremely fragile. The forms of
pollution—ozone, smog, inversion layer, acid rain—are explained. The
human activities that cause pollution are shown. Indicates methods and
costs of air pollution control.

Air Pollution: A Medical Investigation of One Aspect
British Department of Education and Science
Elizabeth House
York Road

London, England
Color, 17 minutes, optical sound, 16mm., 1972.

Shows Saint Bartholomew's Hospital Medical College where scientists study air pollutants, including vehicle exhausts, and their effect on people. Includes blood sampling, respiration, and performance tests.

Air Pollution and Plant Life
National Audio-Visual Center
General Services Administration, Reference Section
8700 Edgeworth Drive
Capitol Heights, MD 20743-3701
Color, 20 minutes, sound, 16mm., 1970.

Surveys air pollution injury to vegetation across the country by common atmospheric pollutants such as sulfur dioxide, oxides of nitrogen, fluorides, and ozone. Environmental and ecological damage is surveyed in Tennessee, Alabama, Florida, West Virginia, Maryland, and California. Suggested for technical audiences in industry, government, and colleges.

Air Pollution: A Trilogy
University of Wisconsin, La Crosse
Film Library Audio-Visual Center
1705 State Street
La Crosse, WI 54601
Color, 7 minutes, sound, 16mm., 1974.

Three films—*Ecomega, Survival,* and *Air*—have been combined into a simple film to present succinct variations on the theme of air pollution by three very different artists.

Air Pollution Dispersion
University of California, Los Angeles
Instructional Media Library
46 Powell Library
Los Angeles, CA 90024
Color, 25 minutes, sound, 16mm., 1976.

Elaborates on most of the factors that determine the way smoke and exhaust gases are dispersed into the atmosphere. The effects of hills and buildings on nearby chimneys and how pollution hazards can be measured in certain atmospheric conditions are discussed.

Air Pollution—Everyone's Problem
Kaiser Steel
300 Lakeside Drive
Oakland, CA 94666
Color, 20 minutes, optical sound, 16mm., 1969.

Presents the story of air pollution, its causes and effects. Explains why the Los Angeles basin is a prime area for smog and shows the efforts of Kaiser Steel's Fontana Plant to control the problem of an industry's pollution.

Air Pollution in the Southcoast Basin—Time Lapse Smog
Audio-Visual Aids
University of California at Riverside
Riverside, CA 92521-4009
Color, 4 minutes, silent, 16mm., 1972.
Uses time lapse photography, compressing 14 hours of the day into four minutes' viewing time. Shows the movement of air pollution into the San Bernardino Mountains of southern California.

Air Pollution: Sweetening the Air
Document Associates
211 E. 43 Street
New York, NY 10017
Color, 24 minutes, sound, 16mm., 1971.
Presents the need for recycling waste products. If the present rate of pollution continues (half a million tons daily is discharged into the atmosphere), life as it now exists will be changed drastically. A project of the University of Toronto is reported in which scientists are studying the microstructure of fly ash in order to find a use for it. Shows other recycling projects.

Air Pollution: Take a Deep, Deadly Breath
American Broadcasting Company, TV
1330 Avenue of the Americas
New York, NY 10019
Color, 54 minutes, sound, 16mm., 1968.
The film surveys ways that have been taken to curb air pollution. Demonstrates why state and local antipollution efforts have been inadequate. Describes the role of the federal government and analyzes proposed solutions to urban pollution problems.

Pollution: Japan's Greatest Challenge
Japan Institute for Social and Economic Affairs
Charles Von Lowenfeldt
1333 Gough Street
San Francisco, CA 94109
Color, 26 minutes, sound, 16mm., 1973.
Illustrates the growing problems associated with pollution and solid waste in Japan and documents efforts to control them. Shows the effects

of smog in Tokyo and other cities as well as the harmful effects on the rural environment. The high altitude effect of jet aircraft exhausts on atmospheric conditions is discussed. The film is optimistic.

Coal and Air Pollution

Coal: Solution or Pollution?
Pennsylvania State University
Audio Visual Services
University Park, PA 16802
Color, 30 minutes, sound, ³/₄″ U-Matic, ¹/₂″ VHS, 1980.

Evaluates the technical problems associated with coal utilization, focusing on the removal of sulfur from coal and weighing the benefits of clear air against the inevitable costs. Raises serious questions of whether we are willing to pay the economic costs of cleaning coal, the environmental costs of using untreated coal, and the social costs of not using coal.

Coal: The Other Energy
National Audio-Visual Center, General Services Administration,
    Reference Section
8700 Edgeworth Drive
Capitol Heights, MD 20743-3701
Color, 15 minutes, sound, 16mm., 1978.

Provides a realistic look at the U.S. coal resources and the problems and possibilities associated with their use now and in the future. Shows the plant technology and processes needed to extract and burn coal without scarring the earth or polluting the atmosphere.

Coal: The Sleeping Giant
University of Colorado
Educational Media Center
Campus Box 379
Boulder, CO 80309
Color, 29 minutes, sound, 16mm., 1979.

There is a great abundance of coal in the United States. The questions are, How do we mine coal with the least effect on the land? How do we clean the coal so that it can be burned without emitting dangerous pollutants into the atmosphere? How do we meet the rising costs of labor, equipment, and delivery systems? How do we turn the coal into a clean liquid product?

Acid Precipitation

Acid from Heaven
National Film Board of Canada
1251 Avenue of the Americas
New York, NY 10020
Color, 30 minutes, sound, 16 ...

... industries and cars and trucks pollute the air with sulfur and nitric oxides. These oxides undergo chemical changes converting them to acid-causing sulfates and nitrates, which fall back to earth, often in the form of acid rain or snow. Acid rain destroys or harms whole lakes by slowly killing the fish and plant life, eats away at stone statues and buildings, and corrodes metal on cars and rooftops.

Acid Precipitation: Particles and Rain
Film Fair Communications
10900 Ventura Boulevard
P.O. Box 1728
Studio City, CA 91604
Color, 16 minutes, sound, Beta, VHS, ³/₄" U-Matic, 1984.

The problems of acid rain are evaluated.

Acid Rain
Time-Life Video
1271 Avenue of the Americas
New York, NY 10020
Color, 57 minutes, sound, Beta, VHS, ³/₄" U-Matic, 1984.

An in-depth analysis of the problems of acid rain in the environment.

Acid Rain
Film Fair Communications
10900 Ventura Boulevard
P.O. Box 1728
Studio City, CA 91604
Color, 17 minutes, sound, Beta, VHS, ³/₄" U-Matic, 1984.

A comprehensive and basic view of the causes and disastrous effects of acid rain on the environment.

Acid Rain
Film Fair Communications
10900 Ventura Boulevard
P.O. Box 1728

Studio City, CA 91604
Color, 16 minutes, sound, 16mm., 1984.

Effects of acid pollutants in the air on vegetation, lakes, and buildings. How our industrial civilization has contributed to atmospheric pollutants. Sulfur dioxide and carbon dioxide control of rain, snow, and ice formations are shown. Methods of overcoming problems and alternative energy sources are discussed.

Acid Rain: Requiem or Recovery?
National Film Board of Canada
1251 Avenue of the Americas
New York, NY 10020
Color, 27 minutes, sound, 16mm., 1982.

Examines the origins, range, and environmental impact of acid rain. Blown by prevailing winds and returned to the earth in the form of dew, rain, or snow, sulfuric and nitric acids enter the earth's life cycle, killing natural plants and animal life and eating away sandstone, marble, limestone, and metals. Graphs and animations illustrate the geographic areas in the United States and Canada most sensitive to acid rain.

Acid Rain: The Choice Is Ours
TV Sports Scene (TVSS Inc.)
5804 Ayrshire Boulevard
Minneapolis, MN 55436
Color, 20 minutes, sound, Beta, VHS, ³/₄" U-Matic, 1982.

Acid precipitation attributed to increased fossil fuel consumption in industrial regions of Europe and North America is examined for the negative effects on life.

Acid Reign
Kinetic Film Enterprises
255 Delaware Avenue
Buffalo, NY 14202
Color, 10 minutes, sound, ³/₄" U-Matic, 1983.

This film, narrated by Ralph Nader, presents the problem of acid rain as another part of the ongoing battle to preserve the environment.

Ozone

The Hole in the Sky
Coronet
108 Wilmot Road

Deerfield, IL 60015
Color, 58 minutes, sound, ½" VHS, 1987.

Discusses the decrease in the protective ozone layer over the South Pole thought to be caused by chlorofluorocarbons. The film considers the consequences of the future reduction of ozone on a world scale.

The Ozone Story
University of Utah Educational M~~~~ ~
207 M~~~~ ~~

Color, 15 minutes, sound, ½" VHS, n.d.

Explains that ozone is one of the most poisonous chemicals known to humans but most necessary to human survival. Presents work performed in York University laboratory and field studies. The production of fluorocarbon and its use, the effect of ultraviolet light in producing suntan and skin cancer, and animated drawings of ozone formation and destruction are presented.

Smog

Smog
Macmillan Films
5547 Ravenswood Avenue
Chicago, IL 60640
Black and white, 8 minutes, sound, 16mm., 1967.

Light but pointed satire on the problems of air pollution in our cities. One of the effects is human blindness. A man-on-the-street interviewer chokes to death on the very smog he dare not see.

Indoor Air Pollution—Smoking

Smoke Screen
Pyramid Film and Video
P.O. Box 1048
Santa Monica, CA 90406
Color/black and white, 5 minutes, sound, 16mm., 1970.

Demonstrates the dangers of smoking without inhaling. Rapidly flashing stills from 40 years of cigarette advertising dissolve into multiple images of smokers, cancer patients, tissue slides, open heart surgery, and news headlines. Synchronized music—beginning with "Smoke Gets in

Your Eyes" and gradually disintegrating into discordant electronic sounds—heightens visual contrasts.

Weather Modification

Tame the Wind
Journal Films
930 Pitner Avenue
Evanston, IL 60202
Color, 28 minutes, sound, 16mm., 1974.

By polluting the air, human activities are inadvertently altering the weather and reducing the quality of life. The film shows how humans are attempting to correct environmental devastation and to improve the weather through weather modification. A number of examples are given. Following a devastating earthquake, a Yugoslavian city is rebuilt so as to optimize ventilation and reduce air pollution potential; in Israel, clouds are seeded in an attempt to stimulate rain; at a Paris airport, fog is dissipated by jet engines. This is a good survey of weather modifications.

Volcanic Pollution

Volcanic Eruptions, Part I
University of Texas
Film Division
Petroleum Extension Service
10100 Burnet Road
Austin, TX 78758
Black and white, 27 minutes, sound, 16mm., 1960.

Volcanoes are described and categorized, followed by examples of different kinds of volcanoes.

Volcanic Eruptions, Part II
University of Texas
Film Division
Petroleum Extension Service
10100 Burnet Road
Austin, TX 78758
Black and white, 26 minutes, sound, 16mm., 1960.

Many different kinds of volcanic eruptions are described.

Volcano
Finley, Stuart
3428 Mansfield Road
Falls Church, VA 22041
Black and white, 17 minutes, sound, 16mm., 1960.

Heat, gases, and pressure as causes of volcanic action are discussed.
Examples shown are Mauna Loa, Fujiyama. Vesuvius, Etna, Paricutin,
Mt. Pelée, and Krakatoa.

Volcanoes, Earthquakes, and Other Earth Movements
Journal Films
930 Pitner Avenue
Evanston, IL 60202
Color, 16 minutes, sound, 16mm., 1981.

Uses live photography, demonstrations, and drawings to study the
causes and effects of volcanoes, and earthquakes. Discusses plate tec-
tonics in terms of lava, ash, plates, faults, and ocean ridges. Looks at the
activity of the Pacific Mountain range of North and South America and
the Atlantic oceanic ridge.

Volcanoes in Action
Encyclopedia Britannica Educational Corporation
425 N. Michigan Avenue
Chicago, IL 60611
Black and white, 11 minutes, sound, 16mm., n.d.

Shows evidence of past volcanic action. Explains landforms produced by
volcanoes and products of volcanoes such as ash and cinder cones.
Discusses the causes and distribution of active volcanoes throughout the
world.

Volcanoes of Mexico
United World Films
221 Park Avenue South
New York, NY 10003
Color, 8 minutes, sound, 16mm., 1965.

Shows the Pacific "girdle of fire" where two-thirds of the world's active
volcanoes are located. Scenes of eruption of Paracutin 1943–1952 are
emphasized.

Glossary

acid precipitation Precipitation that has a pH rating lower than 5.7.

aerosol Suspended material such as dust particles and condensation nuclei in the atmosphere.

allergy Excess sensitivity producing a bodily reaction to certain substances, such as food, pollen, drugs, or heat or cold, which are harmless to most persons.

anions Negative ions.

biota The animal and plant life of a given region or period.

bronchitis An inflammation of the lining membrane of the bronchi or bronchia.

carbon dioxide A carbon oxygen (CO_2), trace element in the atmosphere.

carbon dioxide pools The interchange of carbon dioxide between the atmosphere, biota, and oceanic surfaces.

carbon monoxide A colorless, odorless, very poisonous gas (CO), which burns with a pale-blue flame and is formed when carbon burns with an insufficient supply of air.

cation A positive ion.

clean air Air formed far removed from human habitation and not affected by catastrophic and violent human phenomena.

desulfurization The removal of sulfur from coal.

diatomic oxygen   A molecule of oxygen consisting of two atoms (O_2).

dolomite   A magnesium carbonate ($MgCO_3$) rock.

ecology   The branch of biology that studies the relationships among organisms and their total environment, both animate and inanimate.

ecosystem   The basic ecological unit, made up of a community of organisms interacting with their inanimate environment.

emphysema   A disease of the lungs characterized by a thinning of the lung tissues and a loss of their elasticity.

feedback   A regulation of temperature by a recirculation of energy in the atmosphere.

fission reaction   The conversion of radioactive materials into energy.

flue gas   Gases released in the combustion of a fuel that escape through the stack.

fluidized bed combustion   Burning of coal on a perforated bed containing mineral matter, usually limestone or dolomite and residual ash from previously burned coal, to remove sulfur in the process of combustion.

fossil fuels   Sources of energy formed when dead animals and/or plants were converted into liquids (oils), gases (natural gas), or solids (peat and coal) through geologic time.

genetics   The science of the hereditary and evolutionary similarities and differences of related organisms, as produced by the interaction of the genes.

greenhouse effect   In the greenhouse effect the sun's short wave rays pass through the atmosphere with little change. When they strike a surface the short wave rays are changed to long wave rays releasing energy to heat the atmosphere. As a result the earth receives not only solar radiation but also the longer wave radiation absorbed by the atmosphere and retained.

hydrotreatment   The treatment of high-sulfur oils with hydrogen to remove sulfur.

ionize   To separate into ions.

isotope   Any of two or more forms of the same element having the same atomic number and nearly the same properties but with different atomic weights.

laser radar   A device for amplifying light radiation in which a beam of light is shot through the crystal to emit an intense, direct light beam that is useful in micromachining, surgery, and computer desi~

~~~~~ ~~ ~ point

~~~~~~, measured in degrees on the meridian of the point.

leukemia   A fatal disease of the blood in which there is a pronounced increase in the number of leukocytes. Cancer of the blood.

limestone   A calcium carbonate ($CaCo_3$) rock.

liming   The use of limestone in an acidified lake to reduce the acid content.

limnology   Scientific study of lakes.

line source of pollution   The connection of a number of point sources of emissions into the atmosphere.

melanoma   A tumor, usually malignant, composed of cells containing dark pigment.

mutation   A sudden inheritable change appearing in the offspring of a parent organism owing to an alteration in a gene or chromosome.

neutralization   The reduction of acid in an acidified lake to a neutral condition.

nitric oxide   A colorless, gaseous compound (NO_X).

nuclear energy   Energy produced from radioactive materials.

oxidation   The act or process of uniting with oxygen.

ozone   A form of oxygen (O_3) having three atoms to the molecule, with an odor suggesting that of weak chlorine, produced when an electric spark is passed through oxygen or air and found in the atmosphere in minute quantities after a thunderstorm.

pH scale   A scale measuring acidity or alkalinity of a liquid. Seven indicates an absolutely neutral substance.

photochemical oxidants   Chemical combination of ozone (O_3), nitric oxide (NO_X), and small quantities of peroxyacetyl nitrate (PAN).

photochemical reactions   Hydrocarbons, hydrocarbon derivatives, and nitric oxides combine chemically to produce smog.

photolysis   The decomposition of materials caused by the action of light.

point source of pollution   Single source of emissions into the atmosphere.

precipitation   The results of atmospheric condensation, i.e., snow, sleet, hail, dew, fog, or rainfall.

pyrite   An iron disulfide (FeS_2) with a brass-yellow color and metallic luster that crystallizes in the isometric system; also fool's gold.

radical   An atom or group of atoms containing one or more impaired electrons.

radioactive plume   Waste materials such as iodine-131, cesium-137, and strontium-90 emitted from a nuclear explosion.

radioactivity   The emission of alpha rays, beta rays, or gamma rays in elements such as uranium that undergo spontaneous atomic disintegration.

radon   Naturally occurring gas produced by the radioactive decay of the element radium.

radon isotopes   The products from the decay of radon.

satellite   A small planet revolving around a larger one.

scrubbing   The removal of high-sulfur flue gases in the chimney of the stack with a chemical absorbent such as lime or limestone after combustion.

silicate   A salt of silicic acid.

smog   A mixture or blend of smoke and fog, as seen in highly populated manufacturing and industrial areas.

solar energy   Energy the earth receives from the sun.

stratosphere   A layer of the atmosphere about seven miles above the surface of the earth, between the troposphere and the thermosphere, within which the temperature remains approximately constant.

sulfur dioxide   A heavy suffocating, colorless, water-soluble gas (SO_2).

sulfuric acid   A colorless, oily, and strong corrosive compound (H_2SO_4).

supersonic   Pertaining to or attaining speeds greater than that of sound.

superstack   A smokestack that exceeds 500 feet in height.

tailings   Waste materials from the processing of mineral ores.

toxic metals   Suspended metals such as nickel

ultraviolet radiation   Beyond the end that is violet in the visible spectrum, as of light rays with very short wavelengths.

UV-B flux   The ultraviolet spectrum of wavelengths from 240 to 329 nm that are critical to life.

water vapor   Water in a gaseous condition, especially below the point of boiling, and present in varying quantities in the atmosphere.

yellowcake   The product of the refining of uranium ore into uranium oxide (U_3O_8).

Index